Brain Stem Cells

EXPERIMENTAL BIOLOGY REVIEWS

Series advisors:

D.W. Lawlor
AFRC Institute of Arable Crops Research, Rothamsted Experimental Station, Harpenden, Hertfordshire AL5 2JQ, UK

M. Thorndyke
School of Biological Sciences, Royal Holloway, University of London, Egham, Surrey TW20 0EX, UK

Environmental Stress and Gene Regulation
Sex Determination in Plants
Plant Carbohydrate Biochemistry
Programmed Cell Death in Animals and Plants – SEB Symposium Series Vol. 52
Biomechanics in Animal Behaviour
Cell and Molecular Biology of Wood Formation
Molecular Mechanisms of Metabolic Arrest
Brain Stem Cells – SEB Symposium Series Vol. 53

Forthcoming titles include:
Endocrine Interactions of Insect Parasites and Pathogens
Environment and Animal Development

SOCIETY FOR EXPERIMENTAL BIOLOGY SYMPOSIA

SEB Symposia form a long-standing series of volumes first published in 1947. The series is annual and each publication is a collection of authoritative articles on an aspect of modern experimental Biology. The contributors are all invited and speak on specific topics within the chosen field. Meetings are held annually, in September, over a period of two or three days.

The aims of the Symposium series are to stimulate discussion and communication between scientists of all nationalities, to foster the development of, and research on, modern aspects of plant, animal, and cell biology.

Brain Stem Cells

Edited by

J.A. MIYAN
Department of Biomolecular Sciences, UMIST, Manchester M60 1QD, UK

M. THORNDYKE and P.W. BEESLEY
School of Biological Sciences, University of London, Egham TW20 0EX, UK

C.M. BANNISTER
Department of Neurosurgery, Royal Manchester Children's Hospital,
Manchester M27 4HA, UK

BIOS

First published in 2001

A CIP catalogue record for this book is available from the British Library.

ISBN 1 85996 222 X

BIOS Scientific Publishers Ltd
9 Newtec Place, Magdalen Road, Oxford OX4 1RE, UK.
Tel. +44 (0)1865 726286. Fax. +44 (0)1865 246823
World Wide Web home page: http://www.bios.co.uk/

Distributed exclusively in the United States, its dependent territories, Canada, Mexico, Central and South America, and the Caribbean by Springer-Verlag New York Inc, 175 Fifth Avenue, New York, USA, by arrangement with BIOS Scientific Publishers, Ltd, 9 Newtec Place, Magdalen Road, Oxford OX4 1RE, UK.

Production Editor: Andrea Bosher.
Typeset by Saxon Graphics Ltd, Derby, UK.
Printed by Cromwell Press, Trowbridge, UK.

Contents

Contributors

H. Abraham, Central Electron Microscopic Laboratory, Faculty of Medicine, University of Pécs, Hungary

C.M. Bannister, Department of Biomolecular Sciences, University of Manchester Institute of Science & Technology, PO Box 88, Sackville Street, Manchester M60 1QD, UK and Department of Neurosurgery, Royal Manchester Children's Hospital, Hospital Road, Pendlebury, Manchester M27 4HA, UK

P.W. Beesley, School of Biological Sciences, Royal Holloway, University of London, Egham, Surrey TW20 0EX, UK

W.-C. Chen, School of Biological Sciences, Royal Holloway, University of London, Egham, Surrey TW20 0EX, UK

G. D'Arcangelo, Department of Pediatrics-Neurology, Baylor College of Medicine, The Cain Foundation Laboratories. The Feigin Center, Suite 955, MC 3–6365, 1102 Bates Street, Houston, TX 77030, USA

M. Frotscher, Institute of Anatomy, University of Freiburg, PO Box 111, D-79001 Freiburg, Germany

C.T. Gregg, Genes and Development Research Group, University of Calgary Faculty of Medicine, 3330 Hospital Drive NW, Calgary, Alberta, Canada T2N 4N1

B. Heimrich, Institute of Anatomy, University of Freiburg, PO Box 111, D-79001 Freiburg, Germany

J.A. Knoblich, Research Institute of Molecular Pathology (IMP), Dr. Bohr Gasse 7, A-1030 Vienna, Austria

M. Li, Centre for Genome Research, University of Edinburgh, West Mains Road, King's Buildings, Edinburgh EH9 3JQ, UK

F. Mashayekhi, Department of Biomolecular Sciences, University of Manchester Institute of Science & Technology, PO Box 88, Sackville Street, Manchester M60 1QD, UK

J.A. Miyan, Department of Biomolecular Sciences, University of Manchester Institute of Science & Technology, PO Box 88, Sackville Street, Manchester M60 1QD, UK

M. Patruno, School of Biological Sciences, Royal Holloway, University of London, Egham, Surrey TW20 0EX, UK

D. Price, Department of Biomedical Science, University of Edinburgh, Hugh Robson Building, George Square, Edinburgh EH8 9XD, UK

L. Seress, Institute of Physiology, University Medical School Pecs, Hungary

T. Shingo, Genes and Development Research Group, University of Calgary Faculty of Medicine, 3330 Hospital Drive NW, Calgary, Alberta, Canada T2N 4N1

A. Smith, Centre for Genome Research, University of Edinburgh, West Mains Road, King's Buildings, Edinburgh EH9, 3JQ, UK

M.C. Thorndyke, School of Biological Sciences, Royal Holloway, University of London, Egham, Surrey, TW20 0EX, UK

N. Uchida, StemCells Inc., 3155 Porter Drive, Palo Alto, CA 94304, USA

S. Weiss, Genes and Development Research Group, University of Calgary Faculty of Medicine, 3330 Hospital Drive NW, Calgary, Alberta, Canada T2N 4N1

T. Yagi, Division of Molecular Biology, Institute for Cellular Molecular Biology, Osaka University, Yamadaoka 1–3, Suite 565–0871, Japan, and Laboratory of Neurobiology and Behavioral Genetics, National Institute for Physiological Sciences, Myodaiji, Okazaki 444–8585, Japan

Abbreviations

APP	amyloid precursor protein
BDA	biotinylated dextran amine
BDNF	brain-derived neutrotrophic factor
bFGF	basic fibroblast growth factor
bHLH	basic helix-loop-helix
CNR	cadherin-related neuronal receptor
CNS	central nervous system
CNTF	ciliary neurotrophic
CPP	constitutively proliferating population
CR	Cajal-Retzius
CSF	cerebrospinal fluid
EB	embryoid body
EGF	epidermal growth factor
ES	embryonic stem
FGF	fibroblast growth factor
GFAP	glial fibrillary acidic protein
GMC	ganglion mother cell
HSC	haematopoietic stem cells
ICM	inner cell mass
IGF	insulin-like growth factor
LIF	leukaemia inhibitory factor
MR	magnetic resonance
NFLC	neurofilament light chain
NGF	nerve growth factor
Ngn	neurogenin
NSC	neural stem cells
OB	olfactory bulb
PDGF	platelet-derived growth factor
RA	retinoic acid
RMS	rostral migratory stream
SOP	sensory organ precursor
SVZ	subventricular zone
TCR	T-cell receptor
TH	tyrosine hydroxylase
VLDLR	very low density lipoprotein receptor
VZ	ventricular zone

Preface

The current wide ranging interest in stem cells has spilled over into many aspects of biological research; genetics evolution molecular biology, cell morphology and physiology are but a few of the areas where lines of investigation are being pursued with vigor and enthusiasm. The spin off of the knowledge already gained has had important influences in medical practice, especially in the treatment of hematological disorders requiring bone marrow transplantation.

The appreciation that stem cell therapy might have a wider application, in particular in the central nervous system, has turned a spotlight onto research being carried out on brain stem cells.

This book is the result of deliberations of a diverse group of researchers, all with differing interests in brain stem cells, getting together in Cambridge in the summer of the year 2000 to share knowledge, discuss problems and try to answer questions of mutual interest, to consider the factors influencing brain stem cell function and the progenitor cells they give rise to.

All those taking part are leaders in their fields, all have a clear view of how they expect advances to be made in the future and, more than that, have aspirations that eventually their work will lead to brain stem cell therapy revolutionising the management of many central nervous system disorders for which presently there is little that can be done to improve the outcome.

The glimmers of hope for the future that were glimpsed at this meeting are a spur for those who were present and those who read this book to redouble their efforts and make a reality of that vision of the future.

Jaleel Miyan
Michael Thorndyke
Philip Beesley
Carys Bannister

Introduction

J. Miyan and C. Bannister

At the end of the decade of the brain and the start of the new millennium the Society for Experimental Biology responded to the current exciting challenges in neurobiology by organizing a symposium on brain stem cells. In the tranquil surroundings of Queen's and Downing Colleges, Cambridge, an international line-up of speakers gathered to share recent knowledge and to enthuse about future prospects in the growing field of stem cell biology. The interests of the contributors were divided into three areas of study: stem cell specialists chasing the goal of using these cells to treat a range of neurological conditions, including Parkinson's disease and Alzheimer's disease; biologists interested in the control of proliferation of these cells and their differentiation and migration in normal and diseased states; and those interested in associated phenomena including asymmetric cell division and studies of stem cell systems in organisms other than mammals. Relative to the size and impact of the field, the meeting was small, but successfully covered many important areas of neuronal stem cell biology. The contributions collected in this volume give a concise overview of the field that will be an excellent introduction to the subject for workers wanting to know more and those interested in specializing in the field. Each area covered at the conference has at least one contributed chapter written by a leader in the field.

In the first session we were treated to a succession of master classes from John Parnavelas (London, England), Sam Weiss (Calgary, Canada), Ron McKay (Bethesda, USA) and Clive Svendsen (Cambridge, England) on the origin of cortical neurons, neurogenesis, synaptogenesis and brain repair. Sam Weiss, in spite of having fractured an arm and a leg, was able to give his talk and fully participate in the meeting. Meng Li (Edinburgh, Scotland) presented novel work on deriving neural precursors from embryonic stem cells, and Nobuko Uchida (Sunnyvale, USA) gave a fascinating insight into the similarities of stem cells derived from brain and bone marrow. In her research she employs a technique for isolating neuronal stem cells using the cell surface marker CD133 and has explored the potential impact of this for the rapid isolation of large numbers of such cells using flow cytometry. However, as with bone marrow stem cells, it is important to establish that the cells isolated are indeed stem cells. A single stem cell is sufficient to repopulate the entire blood system of an individual for the rest of his/her life, but this is only possible if the transplanted cell is a pluripotent stem cell and not a later, albeit multipotent progenitor. Clearly, the potential of neuronal stem cells has to be further characterized in this way, especially in the light of recent failures in therapeutic transplantation of stem cells for brain disorders.

There followed a session in which Michael Frotscher (Freiburg, Germany), Gabriella D'Arcangelo (Houston, USA) and Takeshi Yagi (Okazaki, Japan) elaborated on the developmental process involved in the construction of the cerebral cortex and cerebellum involving the reelin signalling cascade (Gleeson and Walsh, 2000). A product of Cajal-Retzius cells, reelin is an extracellular glycoprotein that diffuses through the cortical layers activating specific receptors. These activate intracellular signalling pathways including disabled-1, involved in the migration of neurons along radial glia, and cdk5 and P35, which are implicated in the translocation of neurons from radial glia into the cortex. A malfunction in any of these signalling components leads to severe abnormalities in the cortical architecture and results in neurological dysfunction. Indeed, it was neurological dysfunction that led to the identification of the reeler and scrambler mice in which reelin or disabled-1 expression was abnormal. Further study of the literature reveals a number of other gene mutations that also affect cortical development and underlie neurological conditions such as lissencephaly, holoprosencephaly, anencephaly and hydrocephalus (Rice and Curran, 1999). These conditions involve failure of normal migration of neurons and of normal stratification of neurons within the cortical plate.

Jaleel Miyan (Manchester, England) introduced a new thought into the discussion. The development of the central nervous system involves the formation of the neural tube, followed by the ventricular system, which consists of interconnecting fluid-filled compartments. The generation of the substance of the brain occurs from the germinal matrix lining the ventricular lumen. The cerebrospinal fluid is present throughout the life of an individual. It is actively secreted by the choroid plexus and circulates through the ventricular system, crossing all areas of germinal activity before flowing over the surface of the brain to be absorbed at the superior sagittal venous sinus. This fluid has always presented an enigma. It has been thought to act as a shock absorber, a cooling system, a waste disposal pathway or a flotation system to lift the major blood vessels lying on the under surface of the brain and the brain off the floor of the cranium. Recent evidence suggests it has other, perhaps more important roles in maintaining the physiological homeostasis of the brain environment and as a signalling pathway (Nicholson, 1999). Furthermore, interference to normal flow (as occurs in conditions such as hydrocephalus) results in significant effects on the development of the brain, and particularly the cerebral cortex. Looking at failure of development in conditions such as lissencephaly, the data identifying a link between Cajal-Retzius and meningeal cells, and emerging data from magnetic resonance scans of human fetuses, it seems likely that the CSF is a vital signalling system involved in the development of the cerebral cortex and probably also the rest of the central nervous system.

In the third session, Richard Adams (Bath, England) presented some spectacular video images of neuroblasts undergoing division and aligning their spindles prior to division. The mechanisms and reasons for this remarkable phenomenon were discussed and were further elaborated in the presentation by Juergen Knoblich (Vienna, Austria) who described the alignment processes involved in asymmetric division of neuroblasts during the formation of the *Drosophila* nervous system. How a dividing cell programmes one daughter to go down a particular route and another to go a different one was elaborated in a fascinating manner. Last, but certainly not least, was Mike Thorndyke's (London, England) presentation on the neuronal regeneration capacity of invertebrates. These remarkable creatures hold the secrets of neuronal repair, which are likely to be available in the vertebrates but need to be unlocked. The future may show that invertebrates will provide the key to achieve this in time.

All in all, this brief meeting was filled with exciting new information and concepts that stimulated many informal discussions. It is clear that many other areas of neuronal and brain development, in addition to stem cell biology, are being researched and explored as potential targets for future therapy, to prevent abnormal development and/or to ameliorate neurological disease. We hope that this small volume expresses the infectious enthusiasm of the meeting's contributors, stimulates further research and entices young workers into this important field of research. Many consider the end of the decade of the brain as the start of the millennium of the brain. With the completion also of the genome project, we are entering a post-genomic era in which the identification of novel genes involved in neuronal and brain development will lead to a greater understanding of the processes involved and rapid progress towards the aims expressed in our meeting. One clear message that emerged was to avoid a fixation on one particular aspect of stem cell biology or brain development. There are clearly many aspects yet to be discovered and these may be needed before it will be possible to explore the full potential of brain stem cells for future therapy and to understand why current failures are happening.

Some of the participants at the Cambridge symposium seen together (top) and in discussion mode (below).

References

Gleeson, J.G. and Walsh, A.W. (2000) Neuronal migration disorders: from genetic disease to development mechanisms. *TINS* **23:** 352–358.

Rice, D.S. and Curran, T. (1999) Mutant mice with scrambler brains: understanding the signalling pathways that control cell positioning in the CNS. *Gene Develop.* **13:** 2758–2773.

Nicholson, C. (1999) Signals that go with the flow. *TINS* **22:** 143–145.

1

Neural stem cells of the mammalian forebrain

C.T. Gregg, T. Shingo and S. Weiss

1. Introduction

It has been 35 years since Altman and Das (1965) originally proposed that neurogenesis occurs in the adult mammalian brain. Largely due to the poor availability of neuronal markers, and an inability to repeat these results in humans because of the technological constraints of the time, this finding fell into disrepute for nearly 30 years. In 1992, studies by Reynolds and Weiss revealed the presence of neural stem cells (NSCs) in the embryonic (Reynolds *et al.*, 1992) and adult mammalian forebrain (Reynolds and Weiss, 1992). The subsequent explosion of research in this field has revealed that NSCs probably function in the adult brain to mediate neurogenesis in the olfactory bulb (Lois and Alvarez-Buylla, 1994) and, preliminary studies suggest, in the primate neocortex (Gould *et al.*, 1999). While a large emphasis has been placed on the manipulation of NSCs for clinical purposes in the treatment of neurodegenerative diseases, a growing interest has developed with regard to the role these cells may play in development.

This chapter seeks to review our limited understanding of NSCs in the development of the forebrain. Towards this end, a review of the tools and techniques used to study NSCs will be discussed. Further, we will review the mechanisms, both intrinsic and extrinsic, which may serve to regulate NSC maintenance and proliferation, as well as those involved in the determination of the fate of NSC progeny. Finally, this chapter will discuss adult NSCs and the future trends of this rapidly developing field.

2. What is a neural stem cell?

Stem cells have proved to be notoriously difficult to define. There are many different types of cells, existing in different tissues and at different stages of the developing and adult mammal, which may be potentially defined as stem cells. Each of these different types of stem cells appear to have different functions and different properties ranging from the multipotent stem cells of the brain, to the totipotent and transient embryonic stem cell population of the blastula (Fuchs and Segre, 2000). Perhaps the best definition

Brain Stem Cells, edited by J. Miyan, M. Thorndyke, P.W. Beesley and C. Bannister.

of a stem cell comes from a review by Morrison *et al.* (1997), in which the term is proposed to include all self-renewing precursor cells which have the broadest developmental potential available in a particular tissue at a given time.

In the case of NSC, we are able to distinguish these cells from neural progenitor cells by acknowledging a few key features (reviewed in Weiss *et al.*, 1996a). An NSC is multipotent, and therefore capable of proliferating to produce progenitors which give rise to the three major cell types of the central nervous system (CNS): astrocytes, oligodendrocytes, and neurons (*Figure 1*). Secondly, an NSC is capable of self-renewal through the process of asymmetric cell division. It is important to note that while a process of self-renewal does occur, there is evidence to suggest that NSCs may undergo changes in their abilities to produce different progeny during development (reviewed in Shen *et al.*, 1998). However, it is not well understood if these processes involve intrinsic timing mechanisms or whether they rely upon extrinsic signals from the surrounding environment. Finally, NSCs retain their multipotentiality over time. This last point is an important feature because multipotent progenitor cells which undergo limited self-renewal do exist (Weiss *et al.*, 1996a).

A progenitor cell is defined as a cell that has a limited capacity for self-renewal and produces terminally differentiated cells. The term precursor cell is used to refer to both NSCs and progenitor cells, usually when it is not possible to separate the two populations from one another.

NSCs can be isolated from the entire ventricular neuraxis of the developing and adult mammalian brain (Weiss *et al.*, 1996b). In the developing embryonic brain, *in vitro* studies indicate that NSCs first appear in the primary germinal zone known as the ventricular zone (VZ) (Tropepe *et al.*, 1999), and likely later appear in a secondary germinal zone called the subventricular zone (SVZ), which is just superficial to the

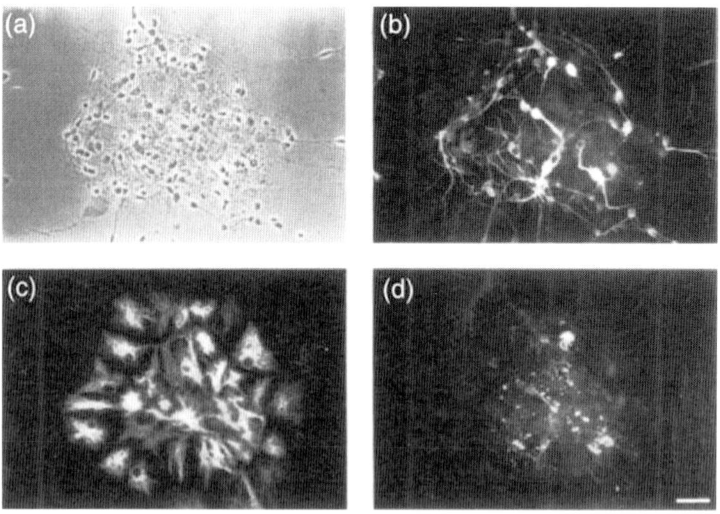

Figure 1. *Neural stem cells proliferate in response to epidermal growth factor (EGF) to produce neurons, astrocytes, and oligodendrocytes. Triple-labelled immunocytochemistry of single EGF-generated neurospheres derived from primary culture (a) with antibodies to MAP-2 (b), GFAP (c), and O4 (d) reveal the presence of neurons, astrocytes, and oligodendrocytes, respectively. Bar, (d), 50 μm. (Figure modified from Reynolds and Weiss, 1996.)*

ventricular zone (Martens *et al.*, 2000). In the adult mammalian brain, NSCs are known to reside as a quiescent population of astrocytes in a cell layer called the subependyma (Doetsch *et al.*, 1999), a remnant of the embryonic SVZ. It is interesting to note that while NSCs do exist in the subependymal region, a recent study has proposed that an adult NSC population may also exist in the single cell layer which lines the lateral ventricles, called the ependyma, a remnant of the embryonic VZ (Johansson *et al.*, 1999).

It is not yet known whether the NSCs that exist in the adult are identical to those that exist in the developing embryo. However, these findings do provide clues to the identity of the NSCs that function embryonically. Indeed, several studies have suggested that a specialized transient glial cell population called radial glia act as an NSC during development (Alvarez-Buylla *et al.*, 1990; Gaiano *et al.*, 2000; Gray and Sanes, 1992; Halliday and Cepko, 1992; Lendahl *et al.*, 1990). Further, it is known that these cells transform into astrocytes in the mature brain (Hunter and Hatten, 1995; Voigt, 1989). Perhaps surprisingly, our *in vivo* understanding of adult NSCs far outweighs our understanding of those found in the embryo. Indeed, we do not even know whether these cells play a role during embryonic development.

3. Fundamental tools for the investigation of forebrain neural stem cells

3.1 In vitro *techniques*

Studies by Reynolds and Weiss have revealed that both embryonic and adult NSCs can be cultured to form clusters of cells known as neurospheres in the presence of a mitogen called epidermal growth factor (EGF), a member of the transforming growth factor alpha (TGFα) superfamily (*Figure 2*; Reynolds *et al.*, 1992; Reynolds and Weiss, 1992; Weiss *et al.*, 1996). It was later found that neurospheres could also be cultured in the presence of basic fibroblast growth factor (FGF-2) and its cofactor, heparin sulfate (Gritti *et al.*, 1996). Much of our knowledge regarding the mechanisms that regulate the production of different NSC-derived progeny has been established by utilizing this *in vitro* technique.

Primary neurospheres are cultured from the germinal zones of the embryo and adult brain. These spheres can be mechanically dissociated into single cell clones and passaged back into culture. Approximately 20% of the EGF-derived cells (Reynolds and Weiss, 1996) and approximately 5% of FGF-2-derived cells (Gritti *et al.*, 1996) will proliferate to form secondary neurospheres. Note that a comparison of these figures may be unfair, as the Reynolds and Weiss (1996) study used embryonic day (E) 14-derived neurospheres, while Gritti *et al.* (1996) used adult-derived neurospheres. Finally, it is important to note that while the neurosphere assay is an important *in vitro* tool, other groups have developed different culture systems for the investigation of neural stem cells as well. An example of such a system involves the use of monolayers attached to the surface of the culture dish, in which cells do not grow as floating spheres, but spread along the floor of the dish.

NSCs and NSC progeny may be differentiated *in vitro* by mitogen withdrawal, or by withdrawing the mitogen and adding a factor which promotes the differentiation of the cells. These cells can then be analysed using immunocytochemistry. This technique has been employed heavily in the search for an immunological marker for NSCs. To

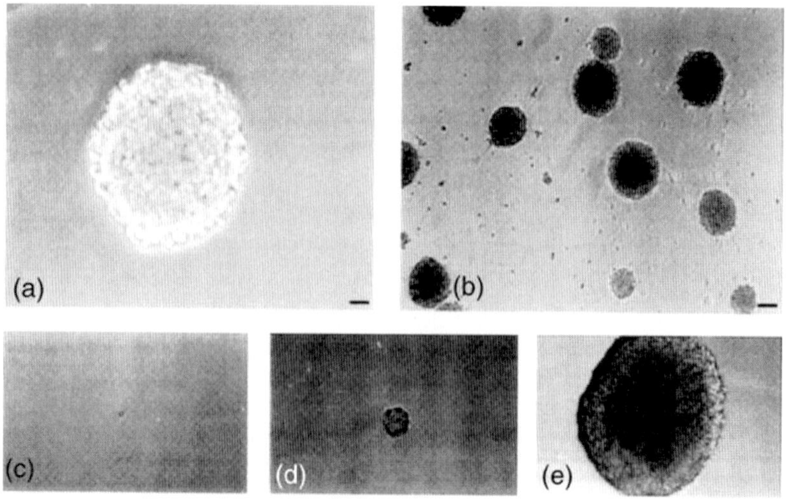

Figure 2. Neural stem cells proliferate in the presence of EGF and fibroblast growth factor 2 (FGF-2) to form neurospheres in vitro. *A single neurosphere (a) may be mechanically dissociated and passaged back into culture, where the dissociated cells will proliferate to give rise to secondary neurospheres (b). Bars: (a), 25 μm; (b), 50 μm. An entire neurosphere is produced from a single neural stem cell (c, d, e). Arrow in (c) identifies the single neural stem cell. (Figure modified from Reynolds and Weiss, 1996.)*

date, this has been an unsuccessful endeavour, but the discovery of such a marker is important for our understanding of what NSCs are and how they arise in development. In addition, the discovery of a surface antigen specific to NSCs may potentially allow for the isolation of these cells, which is important both for their study and for transplantation strategies for neurodegenerative diseases.

While the search for a specific NSC marker has been unsuccessful to date, it has resulted in the identification of several proteins that are expressed by NSCs. These include an RNA-binding protein called Musashi (Kaneko *et al.*, 2000), an intermediate filament protein called Nestin (Lendahl *et al.*, 1990; Reynolds and Weiss, 1992), and glial fibrillary acidic protein (GFAP) (Doetsch *et al.*, 1999). It is significant to understanding the identity of the NSC that these markers are shared by astrocytes, in the case of GFAP and Musashi, and radial glial cells in the case of Nestin (Lendahl *et al.*, 1990). As mentioned previously, radial glial cells are transient and transform into astrocytes later in development (Hunter and Hatten, 1995; Voigt, 1989). The shared expression of these markers suggests that this lineage may be closely related to the NSC lineage.

3.2 In vivo *techniques*

In vivo NSC biology has proved to be a much more difficult topic to address. The use of anti-mitotic drugs (Doetsch *et al.*, 1999), or high doses of tritiated thymidine which kills proliferating cells by intranuclear radiation, have been used to kill the mitotically active precursors of the adult (Morshead *et al.*, 1994) and embryonic (Martens *et al.*, 2000) germinal zone. In the adult, this method removes the more mitotically active progenitor cells of the subependyma, so the only mitotically active cells that remain are the slowly dividing NSCs. This technique has revealed a role for NSCs in the adult as a relatively quiescent cell population which maintains the constitutively dividing progenitors of the rostral migratory stream in the adult forebrain (Doetsch *et al.*, 1999; Morshead *et al.*, 1994). In the embryo, this method has been used to kill embryonic NSCs for the study of NSC proliferation kinetics using a neurosphere-forming assay (Martens *et al.*, 2000).

A second *in vivo* technique employs retroviral lineage tracing, whereby limiting dilutions of retrovirally carried marker genes are injected into the ventricular system of embryonic or adult rat and mouse brains (Craig *et al.*, 1996; Halliday and Cepko, 1992; Morshead *et al.*, 1998). This method depends upon rare infections of precursor cells and the identification of progeny through the undiluted inheritence of the marker gene.

A final technique, which is beginning to be employed at earlier developmental stages with advent of real-time, ultrasound-guided surgeries, involves the transplantation of genetically labelled precursor cells into specific regions of the CNS at very early stages of development (reviewed in Gage, 2000; Gaiano and Fishell, 1998). This method allows researchers to follow the fate of the precursor cell progeny *in vivo*. While a method for isolating an NSC for transplantation has not yet been discovered, the transplantation of genetically labelled neurospheres may reveal a great deal about the role NSCs play during development.

4. Neural stem cells in forebrain development

4.1 *Creating neural stem cells from neuroepithelium*

The induction of the neural plate (reviewed in Weinstein and Hemmati-Brivanlou, 1999) creates a single, relatively homogeneous cell layer which folds to give rise to the neural tube. It is from this homogeneous population of neuroepithelial cells that the first stem cells of the developing brain appear. While this process is not well understood in the mammalian embryo, lessons from *Drosophila* may provide clues as to how NSC specification occurs (Doe *et al.*, 1998; reviewed in Huttner and Brand, 1997).

Approximately one in five cells delaminate from the ventral neuroectoderm of the developing fly embryo to form neuroblasts, the neural stem cells of the fly. These cells undergo repeated asymmetric divisions that renew the neuroblast and produce a smaller progenitor cell called a ganglion mother cell (GMC). Ganglion mother cells are intermediate progenitor cells which undergo a single division to give rise to post-mitotic neuronal siblings, or neuronal and glial siblings (reviewed by Doe and Technau, 1993). The decision as to which cells of the ventral neuroectoderm become neuroblasts and which remain as progenitors of the epidermal tissue is decided by the relative expression levels of proneural and neurogenic genes.

Proneural genes are those which promote a neuroblast fate, while neurogenic genes promote an ectodermoblast fate among the cells of the neuroectoderm (reviewed in Campos-Ortega, 1995). The expression of these genes is regulated by the process of lateral inhibition, which is mediated by a transmembrane receptor called Notch, and its membrane bound ligand, Delta (Campos-Ortega, 1995). As will be discussed later in section 6, homologues of these signalling pathways are also known to play significant roles in the determination of cell fate during neurogenesis and gliogenesis in the mammalian CNS.

Neuroblasts arise from the ventral neuroectoderm through the formation of proneural clusters of cells (*Figure 3*). These proneural clusters are formed by a process called lateral inhibition (for a detailed discussion see reviews by Campos-Ortega, 1995; Huttner and Brand, 1997; Lee, 1997). During this process, the transmembrane receptor, Notch, binds the membrane bound ligand, Delta. This binding event results in the cleavage of the intracellular portion of the Notch receptor by a proteolytic event, which is dependent upon the action of presenilin proteins (Selkoe, 2000). The intracellular domain then complexes with a protein called Suppressor of Hairless and translocates to the nucleus, where it functions as an activating transcription factor for the expression of the basic helix-loop-helix (bHLH) Enhancer-of-Split gene complex (E(Spl)-C). The E(Spl)-C is neurogenic, and inhibits the expression of the proneural genes that compose the achaete-scute complex (AS-C) of bHLH proteins. The inhibition of the AS-C decreases the expression of the transmembrane protein, Delta. This decrease in Notch ligand results in a decrease in Notch signalling in neighbouring cells. Therefore, the expression of the proneural AS-C in neighbouring cells increases as a result of decreased E(spl)-C-mediated repression, and subsequently, this increase in AS-C expression leads to increased Delta ligand expression. Thus, through this process of lateral inhibition, a slight disparity in Notch and Delta expression levels among the cells of the ventral neuroectoderm is amplified to form a heterogeneous cell population. This population then gives rise to the proneural gene-expressing neuroblasts of the developing CNS, and the neurogenic gene-expressing epidermoblasts of the developing epidermal tissue.

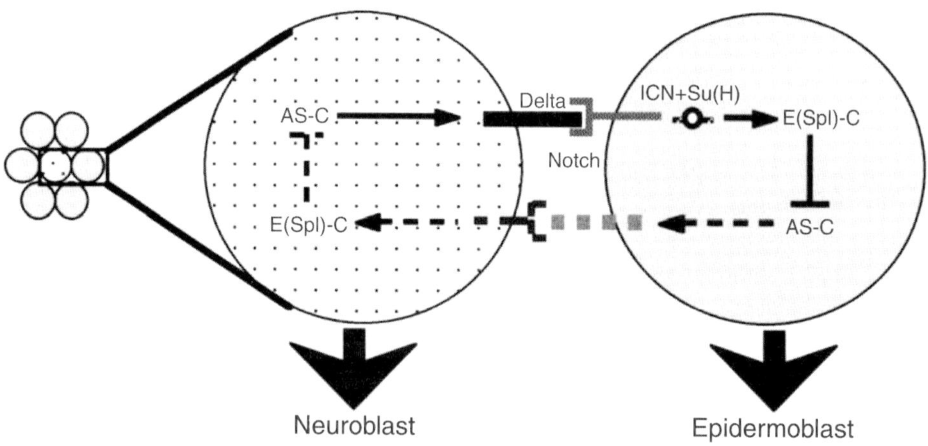

Figure 3. *Notch/Delta signalling mediates lateral inhibition in the ventral neuroectoderm of* Drosophila, *leading to the specification of neuroblasts. Notch activation by binding of the Delta ligand results in the cleavage of the Notch receptor intracellular domain (ICN). The ICN complexes with Suppressor of Hairless (Su(H)) and translocates to the nucleus where it acts as a transcription factor, inducing the expression of the Enhancer-of-Split gene complex (E(Spl)-C). The E(Spl)-C inhibits the expression of the proneural achaete-scute gene complex (AS-C), thereby reducing Delta expression. The reduction in Delta expression results in decreased Notch signalling in neighbouring cells relieving the AS-C from E(Spl)-C-mediated expression. Through this process of lateral inhibition the cells expressing high levels of proneural genes become neuroblasts, while those expressing high levels of neurogenic genes, the Notch effector molecules, become epidermoblasts.*

Homologues of these genes have been identified in several vertebrate systems, including mammals, and are believed to play a major role in generating cellular diversity in the developing CNS (Doe *et al.*, 1998; Huttner and Brand, 1997). However, it remains to be seen whether the mechanisms functioning in *Drosophila* play a role in the specification of NSCs from neuroepithelium in higher organisms.

4.2 *Two different neural stem cells may play different roles in forebrain development*

Two populations of NSCs exist in the developing mammalian brain, which appear to have distinct developmental roles. These cells can be differentiated based upon their responsiveness to the mitogens FGF-2 and EGF (Tropepe *et al.*, 1999). FGF-2-responsive NSCs are the first stem cells to appear, at approximately E8.5, and are believed to reside in the ventricular zone of the developing forebrain (Martens *et al.*, 2000). These cells are then believed to increase in number through symmetric division until approximately E14, after which time they have been suggested to increase their cell cycle time and undergo asymmetric cell division, giving rise to EGF-responsive NSCs (Martens *et al.*, 2000). It has also been suggested in the adult that NSCs exposed to FGF-2 undergo increased levels of neurogenesis, while those exposed to EGF undergo increased levels of gliogenesis (Kuhn *et al.*, 1997). Taken together, these studies support a role for the FGF-2-responsive NSC in primary neurogenesis, which takes place between approximately E9 and E12.5. This is a developmental period during which neurons are born almost exclusively with the exception of radial glial cells.

EGF-responsive stem cells are slowly dividing cells which first appear at the conclusion of primary neurogenesis (Tropepe *et al.*, 1999). A recent study suggests that these cells are the progeny of asymmetrically dividing FGF-2-responsive stem cells in the VZ (Martens *et al.*, 2000). Asymmetric division is understood to occur so that the cleavage plane is parallel to the ventricular surface (Huttner and Brand, 1997), which has been hypothesized to locate EGF-responsive NSCs in the secondary germinal zone, called the SVZ (Martens *et al.*, 2000). This idea is based upon prominent EGF receptor immunostaining in the SVZ, and prominent FGF receptor-1 mRNA levels in the VZ (Eagleson *et al.*, 1996; Wanaka *et al.*, 1991). In culture, EGF-responsive NSCs have been demonstrated to give rise primarily to glial cells, particularly astrocytes (Reynolds and Weiss, 1996). About 10% of the progeny of these NSCs include GABAergic (γ-aminobutyric acid) and substance P immunoreactive neurons (Ahmed *et al.*, 1995). These studies suggest a role for EGF-responsive NSCs during secondary neurogenesis (approximately E13–E19), during which time glia and neurons are born simultaneously; and probably during gliogenesis, which occurs post-natally (approximately P0–P4 in the mouse).

Neurospheres can be attained from the adult subependyma in response to both EGF and FGF-2 (Gritti *et al.*, 1996; Reynolds and Weiss, 1992). However, a recent study by Gritti *et al.* (1999) suggests that a single NSC responds to both of these mitogens in the adult brain. Future research, which focuses on the lineage relationship between adult and embryonic NSCs, may reveal how these different populations come about.

4.3 *Regionalization of neural stem cells in the developing CNS*

A fundamental process that occurs during the development of the CNS involves the regionalization of precursor cells and the subsequent restriction of their progeny to a fate appropriate for a particular region (Fishell, 1997; Jessell and Sanes, 2000; Lumsden and Krumlauf, 1996). This patterning is known to involve both vertical and planar signalling centres. The secretion of the glycoprotein Sonic Hedgehog from the notochord, the prechordal plate and the floorplate mediates ventral vertical signalling after the induction of the neural plate. Dorsal vertical signalling is mediated by members of the TGFβ superfamily, called bone morphogenetic proteins (BMPs), which are secreted from the roofplate and overlying ectoderm. Planar signalling regionalizes the anterior–posterior axis of the developing brain through the secretion of signalling molecules from specialized regions called organizing centres, such as the node, the isthmic organizer and the anterior neural plate. The relationship between these fundamental patterning processes in the developing brain and the generation of progeny from NSCs is very poorly understood.

It may be that the potential of NSCs to generate different cell types in the brain is relatively equal, and that the fate determination of NSC progeny throughout development depends exclusively on vertical and planar signalling from the environment. Alternatively, it may be that NSCs depend partly upon their environment, but become somewhat determined in their abilities to produce particular cell types depending upon their location in the brain. While this is a very poorly understood aspect of NSCs in development, largely because of an inability to isolate these cells for transplantation, there is some evidence to suggest that NSCs are regionalized. The first piece of evidence comes from a study by Weiss *et al.* (1996b) that demonstrates that the mitogen requirements for NSCs of the adult spinal cord are different from those of the adult forebrain when isolated in culture. Further, a recent study by Zappone *et al.* (2000) demonstrates that the transcription factor Sox2 is expressed in embryonic NSCs of the telencephalic VZ when expression of the initializing signal is absent, but not in NSCs of the developing spinal cord. It has also been demonstrated, however, that transplantation of NSCs into ectopic environments causes them to acquire the phenotypic characteristics of the region of transplantation (Gage, 2000; McKay, 1999; Panchision *et al.*, 1998). Thus, it may be that NSCs do undergo some regionalization, but still retain the ability to respond to new extracellular signals.

5. Mechanisms of asymmetric and symmetric cell division

One of the fundamental features of an NSC is its ability to divide asymmetrically. Early in development it is believed that NSCs undergo predominantly symmetric cell division in order to expand the NSC population (Martens *et al.*, 2000). Later in development, asymmetric cell division comes to predominate (Martens *et al.*, 2000). Symmetric division is believed to occur in the ventricular zone when the plane of cleavage is perpendicular to the edge of the ventricular surface. Asymmetric division, on the other hand, occurs when the plane of cleavage is parallel to the edge of the ventricular surface, and results in the unequal distribution of specific proteins to the basally generated progenitor cell and apically renewed stem cell (Huttner and Brand, 1997).

Much of our knowledge of asymmetric cell division comes from studies in *Drosophila* (for a detailed review see Matsuzaki, 2000). Two proteins, called Numb and Prospero, are localized to the basal cortex of dividing neuroblasts. The asymmetric localization of Numb requires a protein called Pon (Partner of Numb) which binds and colocalizes with Numb to the basal cortex of an asymmetrically dividing neuroblast. Numb acts to inhibit Notch signalling in the GMC daughter cell through a direct interaction with the intracellular domain of the Notch receptor allowing for the creation of two distinct cell fates. It is not yet known whether Numb contributes to the difference between neuroblasts and GMCs (Matsuzaki, 2000).

A mammalian Numb homologue (m-Numb) has been demonstrated to be asymmetrically distributed in dividing murine ventricular neural progenitors. In addition, m-Numb has been shown to interact with the mammalian homologue of Notch, Notch1 (Zhong *et al.*, 1996). However, in the mouse, m-Numb is not distributed to the more committed daughter cell, as occurs in *Drosophila*, but appears to remain apically in the NSC. Further, Notch1 is asymmetrically distributed to basally generated cells of the E12.5 ventricular zone, which are presumed to be more committed (Chenn and McConnell, 1995; Zhong *et al.*, 1996). These findings appear difficult to reconcile with the understanding that Notch1 signalling maintains progenitors in an undifferentiated state, and may play a role in the maintenance of the uncommitted stem cell state (reviewed in Morrison *et al.*, 1997). A resolution of this apparent contradiction awaits further research. However, it is interesting to hypothesize that Notch1, a protein involved in the maintenance of the undifferentiated state and, possibly, commitment to glial fate (see section 6.2), may function in the more gliogenic EGF-responsive NSCs, while m-Numb may function in the more neurogenic FGF-2-responsive NSCs, which may divide asymmetrically in the VZ to give rise to EGF-responsive NSCs of the SVZ, as previously discussed.

The Prospero protein is a second protein localized to the basal cortex of dividing *Drosophila* neuroblasts, and is required for correct gene expression in GMC daughter cells. To mediate this function, both the Prospero protein and mRNA are asymmetrically localized to the basal cortex of an asymmetrically dividing neuroblast. The localization of the Prospero protein is mediated by an anchoring protein called Miranda, which has distinct domains that regulate the asymmetric localization of Prospero as well as its subsequent release in the GMC daughter cell, where it acts as a transcription factor. Miranda also basally localizes prospero mRNA through the aid of a double-stranded RNA-binding protein called Staufen which forms a complex with Prospero mRNA.

In order for the proper asymmetric localization of cell fate determining proteins to occur, the mitotic spindle must be orientated parallel to the polar distribution of these determinants (Matsuzaki, 2000). Two proteins act at the apical side of the neuroblast to orient the mitotic spindle. The first is called inscuteable, which coordinates the spindle orientation with the axis of protein localization. The second protein is an apical organizer called Bazooka, which binds directly to inscuteable and localizes it to the apical cortex.

Thus, it is these mechanisms which function to mediate asymmetric cell division in the neuroblasts of *Drosophila*. Further, there is evidence to suggest that many of these same mechanisms are conserved in vertebrates and may function to mediate asymmetric cell division in mammalian NSCs (Chen and McConnell, 1995; Lee, 1997; Zhong *et al.*, 1996).

6. Regulation of neural stem cell-mediated neurogenesis and gliogenesis: extrinsic and intrinsic signals

One of the fundamental areas of NSC study is the investigation of the extrinsic signals and the intrinsic mechanisms that regulate whether a NSC produces neurons or glial cells. Beyond these mechanisms, we are further interested in the processes that determine the production of specific types of neurons and glia.

Previously, the role of EGF, both as a NSC mitogen and as a gliogenic factor, was discussed. In addition, we have reviewed the role of FGF-2 as a second NSC mitogen and as a signalling molecule which promotes neurogenesis. In this section, we will review several other extrinsic factors which also play a role in the choice of a neuronal or glial production (*Figure 4*), and we will also review the intrinsic mechanisms which may act to mediate neurogenic and gliogenic decisions within NSCs.

6.1. *Neurogenesis*

The process of neurogenesis may be mediated by several different signalling molecules, which include brain-derived neurotrophic factor (BDNF), all-*trans* retinoic acid (RA), insulin-like growth factor (IGF-1), and platelet-derived growth factor (PDGF).

BDNF is a member of the neurotrophin family, which includes nerve growth factor (NGF), neurotrophic factor (NT)-3, and NT-4/5. When added to EGF-responsive NSC culture conditions, BDNF enhances the differentiation of NSC neuronal progeny and promotes the generation of GABAergic neurons, but not the survival of neuronal precursor cells (Ahmed *et al.*, 1995). Moreover, *in vivo*, exogenous BDNF has been shown to promote neurogenesis and/or the survival of new neurons in the adult brain (Rasika *et al.*, 1999; Zigova *et al.*, 1998). All-*trans* RA is a derivative of

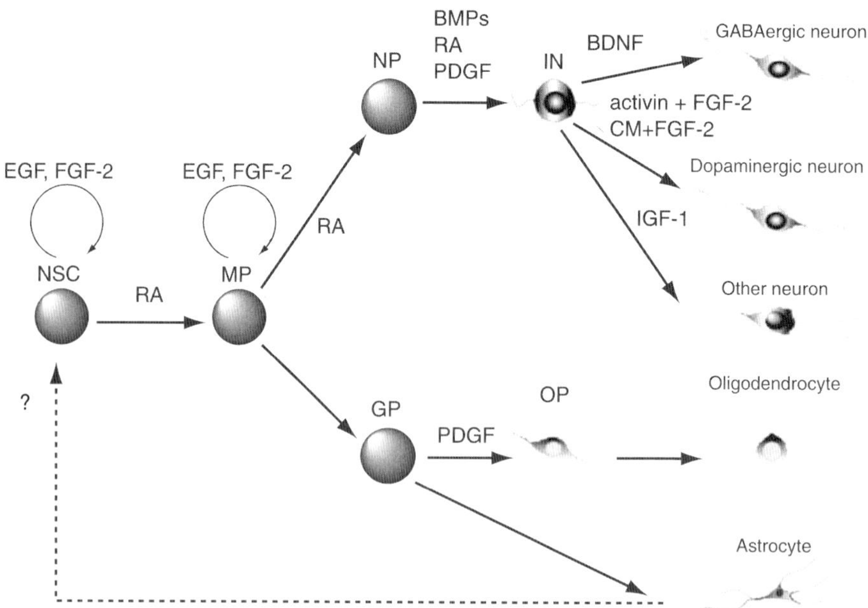

Figure 4. *Extrinsic signalling molecules which regulate the proliferation and self-renewal of neural stem cells (NSCs), and the fate determination of NSC progeny. EGF, epidermal growth factor; FGF-2, fibroblast growth factor-2; PDGF, platelet-derived growth factor; RA, retinoic acid; BMP, bone morphogenetic protein; BDNF, brain-derived neurotrophic factor; IGF-1, insulin-derived growth factor-1; CM, glial conditioned media; MP, multipotent progenitor; NP, neuronal progenitor; IN, immature neuron; GP, glial progenitor; OP, oligodendrocyte progenitor.*

vitamin A and participates in the determination of anterior–posterior patterning of neuraxis and promotes neurogenesis in the developing brain. Wohl and Weiss (1998) suggests that RA can regulate the proliferation of mitotically active EGF-responsive NSC progeny to enhance neurogenesis and astrocyte differentiation *in vitro*.

IGF-1 enhances the proliferation, survival and maturation of sympathetic neuroblasts, and the survival of CNS projection and motor neurons (Neff *et al.*, 1993; Zackenfels *et al.*, 1995). Arsenijevic and Weiss (1998) report that IGF-1 can promote the differentiation of post-mitotic mammalian neuronal precursors derived from EGF-responsive NSCs. Interestingly, this factor also appears to promote the generation of phenotypes other than GABAergic neurons, although these have not yet been characterized. However, another recent paper has successfully identified nonGABAergic neuronal progeny from EGF-responsive NSCs (Daadi and Weiss, 1999). These cells were found to be positive for tyrosine hydroxylase (TH), and were generated using glial cell-conditioned media and exposure to FGF-2. A previous study also found that in the presence of FGF-2, activin induces the generation of TH positive cells from VZ progenitors *in vitro* (Daadi *et al.*, 1998).

The final signalling molecule to be discussed in relation to neurogenesis is PDGF. This factor has been shown to induce neuronal differentiation of cortical NSCs after as little as 2 hours of exposure (Williams, 1997). In addition to neurogenic affects, in the presence of FGF-2, PDGF has been shown to promote the proliferation and migration of oligodendrocyte progenitor cells *in vivo* (McKinnon, 1993). Other factors that are known to influence neurogenesis include corticosteroids, which inhibit neurogenesis and are released from the adrenal medulla during a stress response (Montaron *et al.*, 1999). In addition, recent evidence suggests that neurotransmitters may also play a role in the proliferation of NSCs (Ma *et al.*, 2000) and cortical neurogenesis (Antonopoulos *et al.*, 1997).

While our understanding of the extrinsic signalling molecules that mediate NSC neurogenesis is growing, our understanding of the molecules that act within the cell to mediate neurogenesis is also an active area of study. A recent paper by Shimazaki *et al.* (1999) demonstrated that BDNF and IGF-1 upregulate the POU-III homeodomain containing gene, Brn-4, in striatal precursor cells. Loss of function experiments in this study suggest that the Brn-4 transcription factor mediates the actions of epigenetic signals which induce striatal neuron-precursor differentiation.

Along this same line, investigations of the DNA-binding bHLH family of transcription factors in vertebrates reveal that these proteins are likely major determinants in the choice of the production of either neuronal or glial progeny by NSCs (Lee, 1997). A neuronal fate appears to be regulated by those genes classified as positive regulatory bHLH genes, while a glial fate is regulated by those classed as negative regulatory bHLH genes (Cai *et al.*, 2000; Lee, 1997). The expression of these transcription factors is known to be under the control of Notch signalling (Lee, 1997), and it is likely that many of the previously mentioned extrinsic signalling molecules also regulate the expression of these genes.

Studies indicate that the positive regulatory bHLH genes may be classified as those which promote the determination of a neuronal cell type and those which promote the differentiation of a neuronal cell type (Cai *et al.*, 2000; Lee, 1997). Those that are expressed early in development, between E8.5 and E15.5, promote neuronal determination, or rather the competence to differentiate into a neuron. In the forebrain, these include: the vertebrate achaete-scute homologue, Mash1; and the vertebrate atonal related genes, Neurogenin (Ngn) 1 and Ngn 2 (Fode *et al.*, 2000; Lee, 1997). Misexpression studies of these genes have demonstrated that they are capable of promoting neurogenesis, even at stages during which the process of gliogenesis predominates (Cai *et al.*, 2000; Ferreiro *et al.*, 1994; Lee, 1997). Interestingly, these genes do not appear to be capable of blocking gliogenesis in cerebral cortex, but are capable of blocking this process in the retina (Cai *et al.*, 2000). These results indicate that the regulation pathways of neuronal and glial fate decisions differ in different areas of the CNS.

A recent study by Torii *et al.* (2000) has investigated the role of Mash1, as well as a homeodomain-containing protein called Prox-1, in the proliferation and commitment of NSC fate. The expression of these genes in precursor cell populations which included NSCs, resulted in the inhibition of proliferation and the commitment of cells to neuronal and glial fates. Also, studies of neural crest-derived progenitors reveal that Mash1 specifies commitment to a neuronal lineage and maintains a progenitor in a state in which it is competent to respond to further differentiation signals (Edlund and Jessell, 1999). Interestingly, in a related study by Nakamura *et al.* (2000), a model is suggested in which Hes-1, a mammalian homologue of the Hairy and E(Spl) genes, inhibits multipotent cells from committing to a more restricted neuronal fate. The Hes1 transcription factor is expressed throughout the VZ. Further, it is known to be upregulated by Notch1 signalling, and inhibit the expression of Mash-1 by binding to its promoter sequence (Lee, 1997).

Mash1 mutant mice reveal a role for this gene in the determination of neuronal precursors in the ventral telencephalon – specifically, in the SVZ of medial ganglionic eminence, as well as the sensory neurons of the olfactory epithelium, and peripheral autonomic neurons (MGE) (Casarosa *et al.*, 1999; Harris and Hartenstein, 1999; Horton *et al.*, 1999). Further, Mash1 has been demonstrated to directly activate the mammalian homologue of Delta, Delta-like 1 and 3 (Dll1 and Dll3) (Casarosa *et al.*, 1999), thereby indicating a necessary role for Mash1 in the Notch/Delta lateral inhibition pathway that regulates its expression. Interestingly, Mash1 also appears to indirectly negatively autoregulate (Meredith and Johnson, 2000), a process which may also be mediated by the Hes genes.

Studies have not yet targeted a role for the Ngn genes in NSC regulation. However, it is important to note that Ngn1 and Ngn2 are expressed in a domain of the telencephalon which is dorsal and complementary to the ventral Mash-1 expression domain (Fode *et al.*, 2000). The dorsal expression of these determination genes is required to restrict Mash1 expression to its ventral domain (Fode *et al.*, 2000). It will be interesting to discover the role these genes may play in the commitment of NSCs and NSC progeny.

6.2. *Gliogenesis*

Very little is known about the extrinsic signalling molecules that mediate gliogenesis in the NSCs of the forebrain. Two candidate molecules include leukaemia inhibitory factor (LIF) and ciliary neurotrophic factor (CNTF). Leukaemia inhibitory factor signals by binding to the LIF receptor β/gp130 heterodimer, while CNTF signals by binding to the LIF receptor β/gp130/CNTF receptor alpha heterotrimer. Both of these factors promote the differentiation of oligodendroglial progenitors derived from cortical NSCs (Marmur *et al.*, 1998). Although this field awaits further research to provide clues to other extrinsic signalling molecules that are at work during gliogenesis, recent studies have revealed intrinsic mechanisms that may mediate gliogenesis by NSCs in the forebrain.

While the process of neurogenesis is mediated by the expression of the positive regulatory genes, the process of gliogenesis appears to require the expression of the negative regulatory genes. The negative regulatory genes include the downstream effector molecules of the Notch1 signalling cascade, Hes1 and Hes5 (Lee, 1997; Ohtsuka *et al.*, 1999). As previously described in section 3 (see *Figure 3*), the expression of these genes is regulated by lateral inhibition.

A recent study by Morrison *et al.* (2000) indicates that the transient activation of Notch1, the mammalian Notch homologue, is sufficient to switch neural crest stem cells from the process of neurogenesis to gliogenesis. Further, the specification of glial cell fate in the retina is modulated by Hes5 (Hojo *et al.*, 2000). It has also been demonstrated that the expression of Hes1 and Hes5 inhibits neuronal differentiation (Ohtsuka *et al.*, 1999), most likely through the repression of the positive regulatory determination genes. Again, it is interesting to note that the genes that have been implicated in the maintenance of an undifferentiated state may also commit precursor cells to a glial fate. This suggests further that the identity of a NSC is closely related to glial lineages.

In summary, we have reviewed some of the key extrinsic signalling molecules and cell-intrinsic mechanisms that may regulate the production of neurons and glial cells by NSCs. In the next section, we will switch our focus to the NSCs that exist in the adult forebrain.

7. Adult mammalian forebrain neural stem cells

7.1. *Function and potential*

Most adult mammalian CNS neurons are terminally differentiated and, subsequently, have no capacity for self-replacement after injury or disease. However, the presence of multipotent, self-renewing NSCs in the adult brain has presented new avenues for neuron replacement in damaged regions of the CNS. Adult NSCs are present throughout the entire ventricular neuraxis of the adult brain (Morshead and van der Kooy, 1992; Weiss *et al.*, 1996), and perhaps in the hippocampus (*Figure 5*; Cameron *et al.*, 1993). The adult SVZ of the forebrain contains a mixed population of cells which includes a constitutively proliferating population (CPP), a population of migrating neuroblasts, and a population of slowly proliferating astrocytes which have been demonstrated to act as NSCs (Doetsch *et al.*, 1999; Garcia-Verdugo *et al.*, 1998). The neuroblast population migrates tangentially along the rostral migratory stream (RMS) into the olfactory bulb (OB) to replace dead or dying granule and periglomerular neurons (Lois and Alvarez-Buylla, 1993, 1994; Luskin, 1993). In contrast, glial progenitor cells from the SVZ migrate radially to the striatum, corpus callosum, and cortex (Levison and Goldman, 1993; Levison *et al.*, 1993; Luskin and McDermott, 1994).

In the hippocampus, the CPP is found in a thin lamina between the hilar region and the granule cell layer of the dendate gyrus, that is the subgranular zone (Kempermann *et al.*, 1997; Palmer *et al.*, 1997). However, it is suspect whether or not the dentate gyrus-derived precursor cells are actually NSCs. While these cells can produce the three major cell types of the CNS, they have no obvious NSC characteristics *in vitro* and do not demonstrate the production of region-specific progeny.

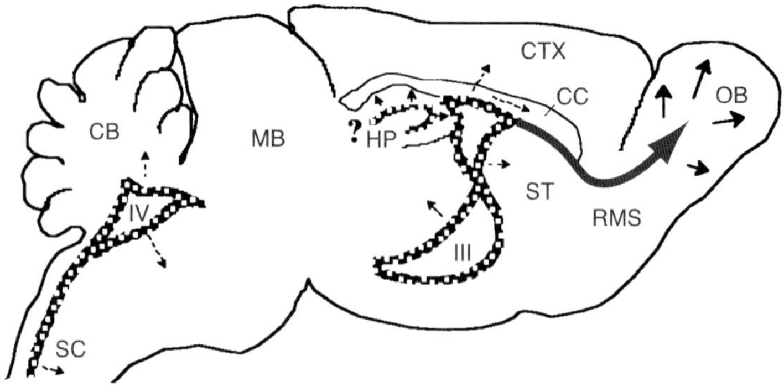

Figure 5. Neural stem cells (NSCs) in the adult mouse brain. NSCs of the adult mouse brain are located in cell layers adjacent to the lateral ventricles (LV), the third ventricle (III), the fourth ventricle (IV), and the central canal of the spinal cord (SC). The NSCs (dotted line) of the third and fourth ventricles function to mediate gliogenesis (dashed arrows) in neighbouring tissues. While neurogenesis (solid arrows) has not been identified in these regions, the NSCs of the LV are known to maintain a constitutively proliferating population of progenitors, which give rise to both neuroblasts and glia. The neuroblasts migrate tangentially in the rostral migratory stream (RMS) to the olfactory bulb (OB), while the glia migrate radially to the corpus callosum (CC), striatum (ST), and cortex (CTX). The hippocampus (HP) is known to undergo neurogenesis in the adult brain to replace the granule cells of the dentate gyrus; however, it is not known whether this process is mediated by NSCs. CB, cerebellum; MB, midbrain.

Spinal cord-derived NSCs can be maintained in culture in the presence of EGF and FGF-2 and reside within the ventricular wall lining the central canal (Weiss *et al.*, 1996b). Interestingly, these cells can transiently proliferate and generate astrocytes in response to injury (Johansson *et al.*, 1999). Further to this point, results from Magavi *et al.* (2000) suggest that endogenous SVZ progenitors may migrate into the cortex and generate new neurons that form synapses following a cortico-thalamic lesion. These results indicate that endogenous NSCs may have some potential to regenerate a damaged brain.

The progeny of adult NSCs is regulated and restricted by their environment, especially by region-specific and defined growth factors. Continuous infusion of exogenous EGF or FGF-2 into the cerebrospinal fluid of the lateral ventricles results in the expansion of the adult NSC population and increased generation of neurons and glia *in vivo* (Craig *et al.*, 1996; Kuhn *et al.*, 1997). Recently, adult NSCs injected into the tail vein of sublethally irradiated mice were found to be capable of generating haematopoietic cells (Bjornson *et al.*, 1999). Moreover, Clarke *et al.* (2000) have demonstrated that adult NSCs contribute to the formation of chimaeric chick and mouse embryos and give rise to cells which comprise the germ layers.

8. Neural stem cells in the human forebrain

Eriksson *et al.* (1998) have demonstrated that a CPP resides in the dentate gyrus of the adult human hippocampus. Also, a study by Pincus *et al.* (1998) found that adult human subependymal cells of the temporal horn have NSC-like characteristics *in vitro*. Further to this point, newborn neurons generated in the SVZ of adult macaques have been found to migrate through the white matter of the cortex to the association neocortex, as well as to the prefrontal, inferior temporal, and posterior parietal cortices (Gould *et al.*, 1999). These studies provide convincing evidence that NSCs are likely to be present and functioning in the adult human brain.

It has also been found that populations containing NSCs can be derived from the human fetus (Gage, 2000). Further, these human fetal stem cells can be transplanted into the rat brain where they differentiate into neurons and glia (Fricker *et al.*, 1999).

9. Future directions

NSC research is still in its infancy, being no more than a decade old and many questions remain to be addressed. With growing evidence for a glial NSC identity it will be interesting to understand the relationship between NSCs and the reactive astrocytes of the adult CNS. These cells are known to proliferate to form glial scars in response to injury, and perhaps they may be manipulated into a multipotent state at the site of injury. Further, an understanding of this relationship will be relevant to cancer research, as gliomas are the most common tumour to occur in the brain. Gene discovery and function studies will also be a primary focus in NSC research as we seek to understand better the mechanisms that regulate stem cell maintenance and cell fate determination. Finally, future research that emphasizes *in vivo* NSC studies in the embryonic and adult brain will better address the function and potential of these cells.

References

Ahmed, S., Reynolds, B.A. and Weiss, S. (1995) BDNF enhances the differentiation but not the survival of CNS stem cell-derived neuronal precursors. *J. Neurosci.* **15:** 5765–5778.

Altman, J. and Das, G.D. (1965) Autoradiographic and histological evidence of postnatal hippocampal neurogenesis in rats. *J. Comp. Neurol.* **124:** 319–335.

Alvarez-Buylla, A., Theelen, M. and Nottebohm, F. (1990) Proliferation 'hot spots' in adult avian ventricular zone reveal radial cell division. *Neuron* **5:** 101–109.

Antonopoulos, J., Pappas, I.S. and Parnavelas, J.G. (1997) Activation of the $GABA_A$ receptor inhibits the proliferative effects of bFGF in cortical progenitor cells. *Eur. J. Neurosci.* **9:** 291–298.

Arsenijevic, Y. and Weiss, S. (1998) Insulin-like growth factor-1 is a differentiation factor for postmitotic CNS stem cell-derived neuronal precursors: distinct actions from those of brain-derived neurotrophic factor. *J. Neurosci.* **15:** 2118–2128.

Bjornson, C.R.R., Rietze, R.L., Reynolds, B.A., Magli, M.C. and Vescovi, A.L. (1999) Turning brain into blood: a hematopoietic fate adopted by adult neural stem cells *in vivo*. *Science* **283:** 534–537.

Cai, L., Morrow, E.M. and Cepko, C.L. (2000) Misexpression of basic helix-loop-helix genes in the murine cerebral cortex affects cell fate choices and neuronal survival. *Development* **127:** 3021–3030.

Cameron, H.A., Woolley, C.S., McEwen, B.S. and Gould, E. (1993) Differentiation of newly born neurons and glia in the dentate gyrus of the adult rat. *Neuroscience* **56:** 337–344.

Campos-Ortega, J.A. (1995) Genetic mechanisms of early neurogenesis in *Drosophila melanogaster*. *Mol. Neurobiol.* **10:** 75–89.

Casarosa, S., Fode, C. and Guillemot, F. (1999) Mash1 regulates neurogenesis in the ventral diencephalon. *Development* **127:** 525–534.

Chenn, A. and McConnell, S.K. (1995) Cleavage orientation and the asymmetric inheritance of Notch1 immunoreactivity in mammalian neurogenesis. *Cell* **82:** 631–641.

Clarke, D.L., Johansson, C.B., Wilbertz, J., Veress, B., Nilsson, E, Karlstrom, H., Lendahl, U. and Frisen, J. (2000) Generalized potential of adult neural stem cells. *Science* **288:** 1660–1663.

Craig, C.G., Tropepe, V., Morshead, C.M., Reynolds, B.A., Weiss, S. and van der Kooy, D. (1996) *In vivo* growth factor expansion of endogenous subependymal neural precursor cell populations in the adult mouse brain. *J. Neurosci.* **16:** 2649–2658.

Daadi, M.M. and Weiss, S. (1999) Generation of tyrosine hydroxylase-producing neurons form precursors of the embryonic and adult forebrain. *J. Neurosci.* **19:** 4484–4497.

Daadi, M.M., Arcellana-Panlilio, M.Y. and Weiss, S. (1998) Activin co-operates with fibroblast growth factor 2 to regulate tyrosine hydroxylase expression in the basal forebrain ventricular zone progenitors. *Neuroscience* **86:** 867–880.

Doe, C.Q. and Technau, G. (1993) Identification and cell lineage of individual neuroblasts in the *Drosophila* CNS. *Trends Neurosci.* **16:** 510–514.

Doe, C.Q., Fuerstneberg, S. and Peng, C. (1998) Neural stem cells: From fly to vertebrates. *J. Neurobiol.* **36:** 111–127.

Doetsch, F., Caille, I., Lim, D.A., Garcia-Verdugo, J.M. and Alvarez-Buylla, A. (1999) Subventricular zone astrocytes are neural stem cells in the adult mammalian brain. *Cell* **97:** 703–716.

Eagleson, K.L., Ferri, R.T. and Levitt, P. (1996) Complementary distribution of collagen type IV and the epidermal growth factor receptor in the rat embryonic telencephalon. *Cereb.Cortex* **6:** 540–549.

Edlund, T. and Jessell, T.M. (1999) Progression from extrinsic to intrinsic signaling in cell fate specification: a view from the nervous system. *Cell* **96:** 211–224.

Eriksson, P.S., Perfilieva, E., Björk-Eriksson, T., Alborn, A-M., Nordborg, C., Peterson, D.A. and Gage, F.H. (1998) Neurogenesis in the adult human hippocampus. *Nat. Med.* **4:** 1313–1317.

Ferreiro, B., Kintner, C., Zimmerman, K., Anderson, D. and Harris, W. (1994) XASH genes promote neurogenesis in Xenopus embryos. *Development* **120:** 3649–3655.

Fishell, G. (1997) Regionalization in the mammalian telencephalon. *Curr. Op. Neurobiol.* **7:** 62–69.

Fode, C., Ma, Q., Casarosa, S., Ang, S., Anderson, D.J. and Guillemot, F. (2000) A role for neural determination genes in specifying the dorsoventral identity of telencephalic neurons. *Genes Dev.* **14:** 67–80.

Fricker, R.A., Carpenter, M.K., Winkler, C., Greco, C., Gates, M.A. and Bjorklund, A. (1999) Site-specific migration and neuronal differentiation of human neural progenitor cells after transplantation in the adult rat brain. *J. Neurosci.* **19:** 5990–6005.

Fuchs, E. and Segre, J.A. (2000) Stem cells: a new lease on life. *Cell* **100:** 143–155.

Gage, F. (2000) Mammalian neural stem cells. *Science* **287:** 1433–1438.

Gaiano, N. and Fishell, G. (1998) Transplantation as a tool to study progenitors within the vertebrate nervous system. *J. Neurobiol.* **36:** 152–161.

Gaiano, N., Nye, J.S. and Fishell, G. (2000) Radial glial identity is promoted by Notch1 signaling in the murine forebrain. *Neuron* **26:** 395–404.

Garcia-Verdugo, J.M., Doetsch, F., Wichterle, H., Lim, D.A. and Alvarez-Buylla, A. (1998) Architecture and cell types of the adult subventricular zone: in search of the stem cells. *J. Neurobiol.* **36:** 234–248.

Gould, E., Reeves, A.J., Graziano, M.S. and Gross, C.G. (1999) Neurogenesis in the neocortex of adult primates. *Science* **286:** 548–552.

Gray, G.E. and Sanes, J.R. (1992) Lineage of radial glia in the chicken optic tectum. *Development* **114:** 271–283.

Gritti, A., Parati, E.A., Cova, L., Frolichsthalm P., Galli, R., Wanke, E. *et al.* (1996) Multipotential stem cells from the adult mouse brain proliferate and self-renew in response to basic fibroblast growth factor. *J. Neurosci.* **16:** 1091–100.

Halliday, A.L. and Cepko, C.L. (1992) Generation and migration of cells in the developing striatum. *Neuron* **9:** 15–26.

Harris, W.A. and Hartenstein, V. (1999) Cellular determination. In: *Fundamental Neuroscience* (eds M.J. Zigmond *et al.*). Academic Press, San Diego, CA, pp. 481–517.

Hojo, M., Ohtsuka, T., Hashimoto, N., Gradwohl, G., Guillemot, F. and Kegyama, R. (2000) Glial cell fate specification modulated by the bHLH gene Hes5 in mouse retina. *Development* **127**: 2515–2522.

Horton, S., Meredith, A., Richardson, J.A. and Johnson, J.E. (1999) Correct coordination of neuronal differentiation events in ventral forebrain requires the bHLH factor MASH1. *Mol.Cell. Neurosci.* **14:** 355–369.

Hunter, K.E. and Hatten, M.E. (1995) Radial glial cell transformation to astrocytes is bidirectional: regulation by a diffusible factor in embryonic forebrain. *Proc. Natl Acad. Sci.USA* **92:** 2061–2065.

Huttner, W.B. and Brand, M. (1997) Asymmetric division and polarity of neuroepithelial cells. *Curr. Op. Neurobiol.* **7**: 29–39.

Jessell, T.M. and Sanes, J.R. (2000) The induction and patterning of the nervous system. In: *Principles of Neural Science,* 4th edn (eds E.R. Kandel, J.R. Schwartz and T.M. Jessell). McGraw-Hill, New York, pp. 1019–1040.

Johansson, C.B., Momma, S., Clarke, D.L., Risling, M., Lendahl, U. and Frisen, J. (1999) Identification of a neural stem cell in the adult mammalian central nervous system. *Cell* **96:** 25–34.

Kaneko, Y., Sakakibara, S., Imai, T., Suzuki, A., Nakamura, Y., Sawamoto, K., Ogawa, Y., Toyama, Y., Miyata, T. and Okano, H. (2000) Musashi1: an evolutionally conserved marker for CNS progenitor cells including neural stem cells. *Dev. Neurosci.* **22**: 139–153.

Kempermann, G., Kuhn, H.G. and Gage, F.H. (1997) Genetic influence on neurogenesis in the dentate gyrus of adult mice. *Proc. Natl Acad. Sci. USA.* **94**: 10409–10414.

Kuhn, H.G., Winkler, J., Kempermann, G., Thal, L.J. and Gage, F.H. (1997) Epidermal growth factor and fibroblast growth factor-2 have different effects on neural progenitors in the adult rat brain. *J. Neurosci.* **17**: 5820–5829

Lee, J.E. (1997) Basic helix-loop-helix genes in neural development. *Curr. Op. Neurobiol.* **7:** 13–20.

Lendahl, U., Zimmerman, L.B. and McKay, R.D. (1990) CNS stem cells express a new class of intermediate filament protein. *Cell* **60:** 585–595.

Levison, S.W. and Goldman, J.E. (1993) Both oligodendrocytes and astrocytes develop from progenitors in the subventricular zone of postnatal rat forebrain. *Neuron* **10:** 201–212.

Levison, S.W., Chuang, C., Abramson, B.J. and Goldman, J.E. (1993) The migrational patterns and developmental fates of glial precursors in the rat subventricular zone are temporally regulated. *Development* **119:** 611–622.

Lois, C. and Alvarez-Buylla, A. (1993) Proliferating subventricular zone cells in the adult mammalian forebrain can differentiate into neurons and glia. *Proc. Natl Acad. Sci. USA* 90: 2074–2077.

Lois, C. and Alvarez-Buylla, A. (1994) Long distance neuronal migration in the adult mammalian brain. *Science* **264:** 1145–1148.

Lumsden A. and Krumlauf, R. (1996) Patterning the vertebrate neuraxis. *Science* **274:** 1109–1115.

Luskin, M. (1993) Restricted proliferation and migration of postnatally generated neurons derived from the forebrain subventricular zone. *Neuron* **11:** 173–189.

Luskin, M.B. and McDermott, K. (1994) Divergent lineages for oligodendrocytes and astrocytes originating in the neonatal forebrain subventricular zone. *Glia* **11:** 211–226.

Ma, W., Maric, D., Li, B.S., Hu, Q., Andreadis, J.D., Grant G.M. ***et al.*** (2000) Acetylcholine stimulates cortical precursor cell proliferation *in vitro* via muscarinic receptor activation and MAP kinase phosphorylation. *Eur. J. Neurosci.* **12:** 1227–1240.

Magavi, S.S., Leavitt, B.R. and Macklis, J.D. (2000) Induction of neurogenesis in the neocortex of adult mice. *Nature* **405:** 951–955.

Marmur, R., Kessler, J.A., Zhu, G., Gokhan, S. and Mehler, M.F. (1998) Differentiation of oligodendroglial progenitors derived from cortical multipotent cells requires extrinsic signals including activation of gp130/LIFbeta receptors. *J. Neurosci.* **18:** 9800–9811.

Martens D., Tropepe, V. and van der Kooy, D. (2000) Separate proliferation kinetics of fibroblast growth factor-responsive and epidermal growth factor-responsive neural stem cells within the embryonic forebrain germinal zone. *J. Neurosci.* **20:** 1085–1095.

Matsuzaki, F. (2000) Asymmetric division of *Drosophila* neural stem cells: a basis for neural diversity. *Curr. Op. Neurobiol.* **10:** 38–44.

McKay, R. (1999) Brain stem cells change their identity. *Nature Med.* **5:** 261–262.

McKinnon, R.D., Smith, C., Behar, T., Smith, T. and Dubois-Dalcq, M. (1993) Distinct effects of FGF-2 and PDGF on oligodendrocyte progenitor cells. *Glia* **7:** 245–254.

Mehler, M.F., Mabie, P.C., Zhu, G., Gokhan, S. and Kessler, J.A. (2000) Developmental changes in progenitor cell responsiveness to bone morphogenetic proteins differentially modulate progressive CNS lineage fate. *Dev. Neurosci.* **22:** 74–85.

Meredith, A. and Johnson, J.E. (2000) Negative autoregulation of Mash1 expression in CNS development. *Dev. Biol.* **222:** 336–346.

Montaron, M.F., Petry, K.G., Rodriguez, J.J., Marinelli, M., Aurousseau, C., Rougon, G., Le Moal, M. and Abrous, D.N. (1999) Adrenalectomy increases neurogenesis but not PSA-NCAM expression in aged dentate gyrus. *Eur. J. Neurosci.* **11:** 1479–1485.

Morrison, S.J., Shah, N.M. and Anderson, D. (1997) Regulatory mechanisms in stem cell biology. *Cell* **88:** 287–298.

Morrison, S.J., Perez, S.E., Qiao, Z., Verdi, J.M., Hicks, C., Weinmaster, G. and Anderson, D.J. (2000) Transient Notch activation initiates an irreversible switch from neurogenesis to gliogenesis by neural crest stem cells. *Cell* **101:** 499–510.

Morshead, C.M. and van der Kooy, D. (1992) Postmitotic death is the fate of constitutively proliferating cells in the subependymal layer of the adult brain. *J. Neurosci.* **12:** 249–256.

Morshead, C.M., Reynolds, B.A., Craig, C.G., McBurney, B.A., Staines, W.A., Morassutti, D., Weiss, S. and van der Kooy, D. (1994) Neural stem cells in the adult mammalian forebrain: A relatively quiescent subpopulation of subependymal cells. *Neuron* **13:** 1071–1082.

Morshead, C.M., Craig, C.G. and van der Kooy, D. (1998) *In vivo* clonal analyses reveal the properties of endogenous neural stem cell proliferation in the adult mammalian forebrain. *Development* **125:** 2251–2261.

Nakamura, Y., Sakakibara, S., Miyata, Y., Ogawa, M., Shimazaki, T., Weiss, S., Kageyama, R. and Okano, H. (2000) The bHLH gene Hes1 as a repressor of the neuronal commitment of CNS stem cells. *J. Neurosci.* **20:** 283–293.

Neff, N.T., Prevette, D., Houenou, L.J., Lewis, M.E., Glicksman, M.A., Yin, Q.W., and Oppenheim, R.W. (1993) Insulin-like growth factors: putative muscle-derived trophic agents that promote motoneuron survival. *J. Neurobiol.* **24:** 1578–1588.

Ohtsuka, T., Ishibashi, M., Gradwohl, G., Nakanishi, S., Guillemot, F. and Kageyama, R. (1999) Hes1 and Hes5 as Notch effectors in mammalian neuronal differentiation. *EMBO J.* **18:** 2196–2207.

Palmer, T.D., Takahashi, J. and Gage, F.H. (1997) The adult rat hippocampus contains primordial neural stem cells. *Mol. Cell. Neurosci.* **8:** 389–404

Panchision D., Hazel T. and McKay R. (1998) Plasticity and stem cells in the vertebrate nervous system. *Curr. Op. Cell Biol.* **10:** 727–733.

Pincus, D.W., Keyoung, H.M., Harrison-Restelli, C., Goodman, R.R., Fraser, R.A.R., Edgar, M., Sakakibara, S., Okano, H., Nedergaard, M. and Goldman, S.A. (1998) Fibroblast growth factor-2/brain-derived neurotrophic factor-associated maturation of new neurons generated from adult human subependymal cells. *Ann. Neurol.* **43:** 576–585.

Rasika, S., Alvarez-Buylla, A. and Mottebohm, F. (1999) BDNF mediates the effects of testosterone on the survival of new neurons in an adult brain. *Neuron* **22:** 55–62.

Reynolds, B.A. and Weiss, S. (1992) Generation of neurons and astrocytes from isolated cells of the adult mammalian central nervous system. *Science* **27:** 1707–1710.

Reynolds, B.A. and Weiss, S. (1996) Clonal and population analyses demonstrate that an EGF responsive mammalian embryonic CNS precursor is a stem cell. *Dev. Biol.* **175:** 1–13.

Reynolds, B.A., Tetzlaff, W. and Weiss, S. (1992) A multipotent EGF responsive striatal embryonic progenitor cell produces neurons and astrocytes. *J. Neurosci.* **12:** 4565–4574.

Selkoe, D.J. (2000) Notch and presenilins in vertebrates and invertebrates: implications for neuronal development and degeneration. *Curr. Op. Neurobiol.* **10:** 50–57.

Shen, Q., Qian, X., Capela, A. and Temple, S. (1998) Stem cells in the embryonic cerebral cortex: their role in histogenesis and patterning. *J. Neurobiol.* **36:** 162–174.

Shimazaki, T., Arsenijevic, Y., Ryan, A.K., Rosenfeld, M.G. and Weiss, S. (1999) A role for the POU-III transcription factor Brn-4 in the regulation of striatal neuron precursor differentiation. *EMBO J.* **18:** 444–456.

Torii, M., Marsuzaki, F., Osumi, N., Kaibuchi, K., Nakamura, S., Casarosa, S., Guillemot, F. and Nakafuku, M. (1999) Transcription factors Mash-1 and Prox-1 delineate early steps in differentiation of neural stem cells in the developing central nervous system. *Development* **126:** 443–456.

Tropepe, V., Sibilis, M., Ciruna, B.G., Rossant, J., Wagner, E.F. and van der Kooy, D. (1999) Distinct neural stem cells proliferate in response to EGF and FGF in the developing mouse telencephalon. *Dev. Biol.* **208:** 166–188.

Voigt, T. (1989) Development of glial cells in the cerebral wall of ferrets: direct tracing of their transformation from radial glia into astrocytes. *J. Comp. Neurol.* **289:** 74–88.

Wanaka, A., Milbrandt, J. and Johnson Jr, E.M. (1991) Expression of FGF receptor gene in rat development. *Development* **111:** 455–468.

Weinstein, D.C. and Hemmati-Brivanlou, A. (1999) Neural induction. *Ann. Rev. Cell. Dev. Biol.* **15:** 411–433.

Weiss, S., Reynolds, B.A., Vescovi, A.L., Morshead, C., Craig, C.G. and van der Kooy, D. (1996a) Is there a neural stem cell in the mammalian forebrain? *Trends Neurosci.* **19:** 387–393.

Weiss, S., Dunne C., Hewson, J., Wohl, C., Wheatley, M., Peterson, A. and Reynolds, B.A. (1996b) Multipotent CNS stem cells are present in the adult mammalian spinal cord and ventricular axis. *J. Neurosci.* **16:** 7599–7609.

Williams, B.P., Park, J.K., Alberta, J.A., Muhlebach, S.G., Hwang, G.Y., Roberts, T.M. and Stiles, C.D. (1997) A PDGF-regulated immediate early gene response initiates neuronal differentiation in ventricular zone progenitor cells. *Neuron* **18:** 553–562.

Wohl, C.A. and Weiss, S. (1998) Retinoic acid enhances neuronal proliferation and astroglial differentiation in cultures of CNS stem cell-derived precursors. *J. Neurobiol.* **37:** 281–290.

Zackenfels, K., Oppenheim, R.W. and Rohrer, H. (1995) Evidence for an important role of IGF-I and IGF-II for the early development of chick sympathetic neurons. *Neuron* **14:** 731–741.

Zappone, M.V., Galli, R., Catena, R., Meani, N., De Biasi, S., Mattel, E. *et al.* (2000) Sox2 regulatory sequences direct expression of a *beta*-geo transgene to telencephalic neural stem cells and precursors of the mouse embryo, revealing regionalization of gene expression in CNS stem cells. *Development* **127:** 2367–2382.

Zhong, W., Feder, J.N., Ming-Ming, J., Yeh Jan, L. and Nung Jan, Y. (1996) Asymmetric localization of a mammalian Numb homologue during mouse cortical neurogenesis. *Neuron* **17:** 43–53.

Zhu, G., Mehler, M.F., Zhao, J., Yung, S.Y. and Kessler, J.A. (1999) Sonic hedgehog and BMP2 exert opposing actions on proliferation and differentiation of embryonic neural progenitor cells. *Dev. Biol.* **215:** 118–129.

Zigova, T.P., Wiegand, S.J. and Luskin, M.B. (1998) Intraventricular administration of BDNF increases the number of newly generated neurons in the adult olfactory bulb. *Mol. Cell. Neurosci.* **11:** 234–245.

2

Similarity and differences between the haemopoietic and brain stem cell system

N. Uchida

Haemopoiesis and neurogenesis proceed from primitive stem cells, to multipotential progenitors, to lineage-committed progenitors, finally differentiating into functionally mature cells. Stem cells are defined as cells that possess the capacity upon cell division to give rise to a daughter cell capable of self-renewal and the other daughter cell that gives rise to a multi-potential progenitor cell. To characterize both haemopoiesis and neurogenesis it is necessary to develop *in vitro* clonogeneic assays that demonstrate self-renewal and multi-potential differentiation of stem cells.

Characterization of murine haemopoietic stem cells (HSC) was achieved by isolating candidate cells based either on surface markers (Baum *et al.*, 1992; Spangrude *et al.*, 1988) metabolic properties (Bertoncello *et al.*, 1985; Visser *et al.*, 1984, 1990) or cell size and density (Jones *et al.*, 1990; Orlic *et al.*, 1994). Murine HSCs are Thy1.1^{lo}, Sca-1+ (Spangrude *et al.*, 1988), c-kit+ (Ikuta and Weissman, 1992), and mature blood lineage marker negative/low (Spangrude *et al.*, 1988). Self-renewal and differentiation capacity of murine HSCs can be evaluated in various clonogeneic assays. The first assay, the spleen colony assay, for the murine haemopoietic system was developed by Till, McCulloch, Wu, Becker and Siminovitch in the 1960s (McCulloch and Till, 1963; Siminovitch *et al.*, 1963; Till and McCulloch, 1961; Wu *et al.*, 1968). Transplanting graded numbers of bone marrow cells into lethally conditioned mice, progenitor cells could be enumerated by the formation of clonogenic spleen colonies, which contained myeloid, erythroid and megakaryocytic precursor cells (multi-lineage differentiation). Stem cell frequency was determined by formation of spleen colonies upon secondary transplantation (self-renewal) (Wu *et al.*, 1967, 1968). The differentiation potential of HSCs for T-lymphocytes can be measured by thymus colony forming assay (Ezine *et al.*, 1984). *In vitro* clonogenic colonies include methylcellulose for myeloerythroid or multi-potential blast colonies (Bradley and Metcalf, 1966; Pluznik and Sachs, 1965), Dexter cultures for myeloerythroid cells (Dexter *et al.*, 1977; Ploemacher *et al.*, 1991) and Whitlock-Witte culture for B-cell lineage (Müller-Sieburg *et al.*, 1986).

Brain Stem Cells, edited by J. Miyan, M. Thorndyke, P.W. Beesley and C. Bannister.

Isolation of human HSC was also achieved based on cell-surface markers. Human HSCs express CD34 (Andrews *et al.*, 1986; Civin *et al.*, 1984), Thy-1 (Baum *et al.*, 1992), and CD133 (Yin *et al.*, 1997) while absent for mature lineage markers (*Figure 1*). *In vitro* clonogenic assays such as long-term culture initiating cells (LTC-IC) (Sotherland *et al.*, 1990) and cobblestone-area forming cell (CAFC) measure primitive stem-cell frequency (Weilbaecher *et al.*, 1991). *In vivo* models for human HSC were developed using xenogeneic transplantation of human cells into immunodeficient mice such as SCID (Kamel-Reid and Dick, 1988; McCune *et al.*, 1988), beige/nude/xid (Nolta *et al.*, 1996) and NOD/SCID mice (Larochelle *et al.*, 1996) recipients. Human HSCs have been tested for homing, expansion and differentiation in an *in utero* sheep transplant model (Uchida *et al.*, 1996; Zanjani *et al.*, 1992). Recently, human clinical studies have shown that transplantation of highly purified stem cells alone is sufficient for rapid and sustained haematopoietic recovery in conditioned recipients (Michallet *et al.*, 2000; Negrin *et al.*, 2000).

In vitro and *in vivo* self-renewal and differentiation potential of HSC has been well established; however, *ex vivo* expansion of self-renewing cells has not been achieved. Both mouse and human HSCs cultured in the presence of a cytokine cocktail, such as IL-3, IL-6, G-CSF and SLF, do not increase the number of primitive stem cells when subsequently tested in various clonogenic assays or *in vivo* transplantation models (reviewed in Weissman and Uchida, 2000).

In contrast, neural stem cells (NSC) can be expanded in long-term cultures with defined cytokines in serum-free conditions (Auerbach *et al.*, 2000; Carpenter *et al.*, 1999; Gage *et al.*, 1995; Lendahl *et al.*, 1990; Ray *et al.*, 1993; Reynolds and Weiss, 1992; Yandava *et al.*, 1999). Rodent neural stem/progenitors have been expanded *in vitro* as a monolayer of cells (Ray *et al.*, 1993) or as neurosphere cultures (Reynolds and Weiss, 1992). These stem/progenitor cells express the neuroepithelial stem cell protein nestin (an intermediate filament protein) (Lendahl *et al.*, 1990) and maintain the capacity of multi-lineage differentiate *in vitro* as well as *in vivo* (Gage *et al.*, 1995). Carpenter *et al.* (1999) optimized conditions to culture non-genetically modified human neural stem/progenitor cells as neurospheres. These neurospheres can expand long-term in a

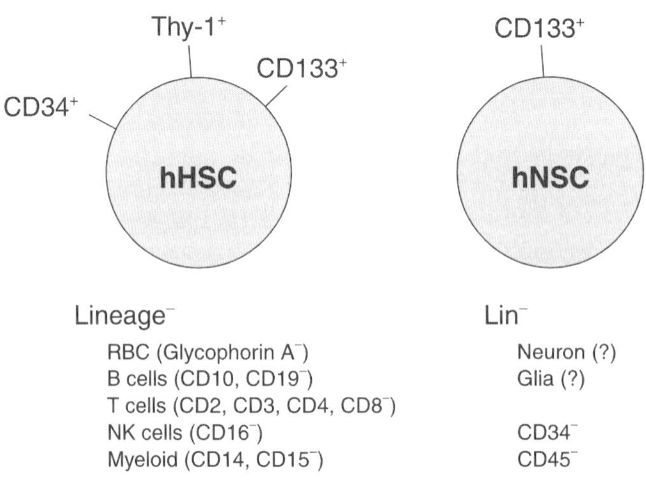

Figure 1. *Searching for CNS stem cell markers.*

serum-free media supplemented with LIF, b-FGF and EGF. When placed under conditions to induce their differentiation, they gradually lose nestin protein expression and differentiate into neurons, astrocytes and oligodendrocytes (Carpenter *et al.*, 1999). NSC can be immortalized with oncogene (e.g., *v-myc*) to allow them to expand indefinitely (Auerbach et al., 2000; Yandava *et al.*, 1999). Human neurospheres passaged by dissociating cells with collagenase instead of conventional trituration resulted in over a 3 log increase in cell numbers after six passages, with an approximate doubling time of 4 days (*Figure 2*).

In contrast to HSC characterization, prospective isolation of NSC was not achieved until recently. Morrison and Anderson isolated rat $p75^+$ P_0^- cells as peripheral nervous system stem cells (PNS-SC) (Morrison *et al.*, 1999). For mouse CNS stem cells (CNS-SC) Johansson and Frisen isolated mouse ependymal cells, which have NSC characteristics, by selecting cells with notch anti-serum from adult brain (Johansson *et al.*, 1999). Doestch and Alvarez-Buylla showed that cells in the subventricular zone, which express the astrocyte marker GFAP, display NSC property (Doetsch *et al.*, 1999).

Our goal is to characterize a population of human CNS-SC from fetal brain based on cell-surface markers and their direct isolation by flow cytometry. To characterize human CNS-SC, we need to establish (i) the long-term *ex vivo* expansion of human CNS-SC; (ii) direct isolation of human CNS-SC from fetal brain based on cell surface markers; (iii) a clongenic assay to evaluate neurosphere-initiating cell (NS-IC); and (iv) xenogeneic transplantation of human CNS-SC into an animal model.

The current markers available for delineating the neurogenic hierarchy are antibodies to either cytoplasmic or nuclear antigens. These include antibodies to nestin, to the neuronal lineage (β-tubulin III, Map2ab, NeuN, etc.) and to the glial lineage (GFAP) (*Figure 3*). To isolate human NSCs, we sought to identify a panel of antibodies reactive to cell surface markers on NSC.

Based on phenotypic identification of mouse haematopoietic stem cells (Ikuta and Weissman, 1992; Spangrude *et al.*, 1988), the isolation of human NSCs requires markers on stem/progenitor cells as well as marker absent on these cells, but present on non-stem cell populations (i.e. mature lineage markers or differentiation antigen). We have

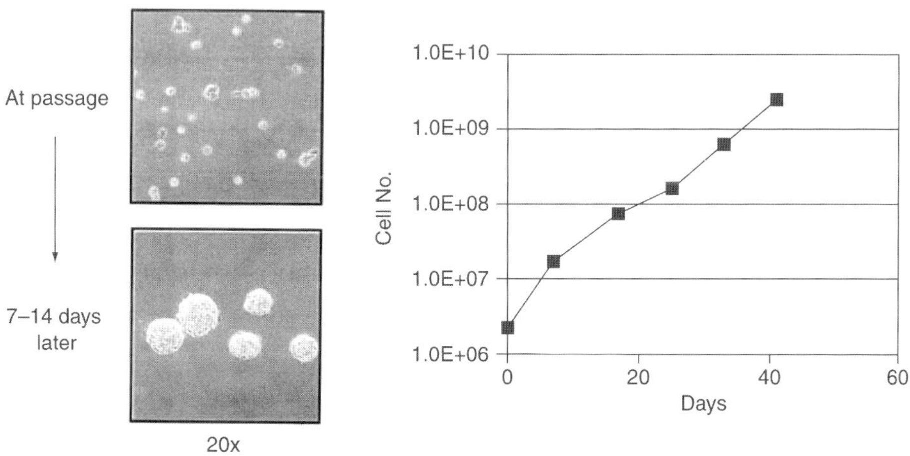

Figure 2. In vitro *proliferation of human neurosphere cells.*

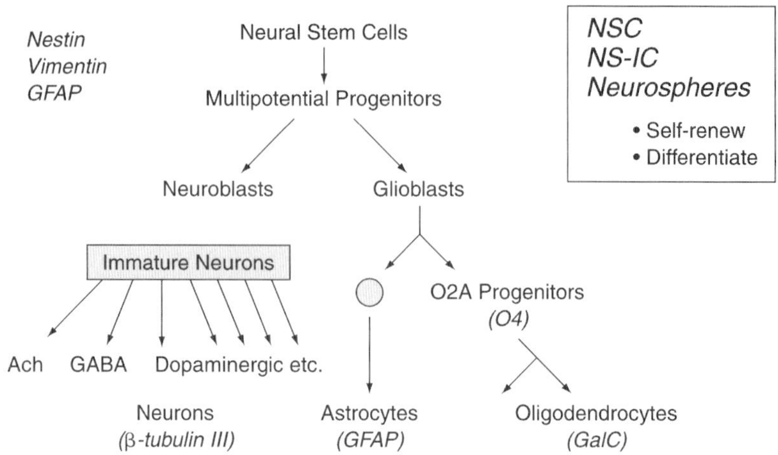

Figure 3. *Neurogenesis.*

screened over 50 known cell surface markers. In addition, we immunized mice with enzyme-dissociated human fetal brains to generate approximately 2000 hybridomas that produced MAbs that were reactive to primary fetal brain tissue or long-term cultured neurosphere cells by flow cytometric analysis. Selected MAbs (approximately 180) were characterized further by immuno-histochemical analysis on developing human fetal spinal cord and brain sections. We have assumed that NSCs from fetal brain can proliferate and self-renew *in vitro* to establish long-term neurosphere cultures. We have used the term neurosphere-initiating cell (NS-IC) to designate a primitive cell that resides in fetal brain and is responsible for initiating neurosphere cultures. To test this hypothesis, it was necessary to identify markers that separate brain cells into two fractions, one enriched with NS-IC activity and the other essentially devoid of activity.

We examined whether human NSC share cell-surface markers previously identified on HSC. Neurosphere cells did not express CD45 and CD34 present on haematopoietic stem/progenitor cells (*Figure 4*). The majority of fetal brain cells expressed high levels of Thy-1; therefore, Thy-1 could not be used to isolate a rare subset of cell population. CD133, originally known as AC133, is a glycosylated 5-transmembrane cell-surface antigen, with molecular weight 120 kDa, which has been shown to be expressed on human HSC (Miraglia *et al.*, 1997; Yin *et al.*, 1997). CD133 is expressed on less than 5% of fetal brain cells from gestational week 16–20, while staining over 90% of long-term neurosphere cells (*Figure 4*). Immunohistological staining revealed that CD133 and 5E12 were expressed in the ventricular zone of the developing brain and spinal cord (10–12 gestational weeks). To characterize a rare subset of fetal brain cells for NS-IC, the cell populations $CD133^+$ $CD34^-$ $CD45^-$ ($CD133^+$) and $CD133^-$ $CD34^-$ $CD45^-$ ($CD133^-$) were separated into highly pure populations. The $CD133^+$ sorted cells proliferated to form neurospheres (*Figure 5*), while $CD133^-$ sorted cells failed to initiate neurospheres. The neurosphere cells derived from expanded $CD133^+$ cells expressed nestin and differentiated into both neural and glial lineages following exposure to differentiation conditions. To evaluate the frequency of NS-IC within a population, we established a clonogenic assay to quantitate NS-IC by limiting dilution analysis. In this assay, the $CD133^+$ subset was

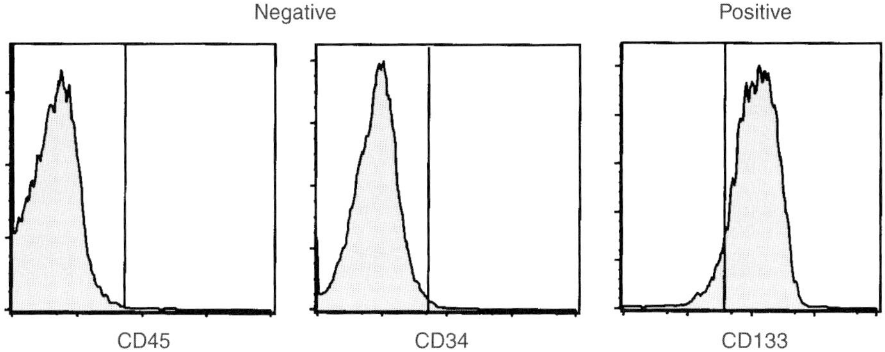

Figure 4. Define cell surface markers on long-term neurosphere cells

enriched for NS-IC activity 23-fold, virtually corresponding to the frequency of NS-IC activity from fetal brain. NS-IC activity was depleted in the $CD133^-$ subset. To address whether a single neurosphere can be derived from a single NSC, $CD133^+$ cells were sorted by flow cytometry and plated as single cells by an automated cell deposition unit (ACDU). Neurosphere cultures were established from single $CD133^+$ cells that expanded in culture and, furthermore, these neurospheres were capable of re-initiating new neurospheres, demonstrating that these cells were self-renewing.

We tested $CD133^+$ sorted/expanded human neurosphere cells in the neonatal NOD-SCID mice xenogeneic transplantation model. Previously, Fricker and Bjorklund and colleagues demonstrated that human neurosphere cells transplanted into immunosuppressed adult rats survive, migrate and differentiate into mature neurons specific for the region of integration (Fricker *et al.*, 1999). Cells were transplanted into the lateral

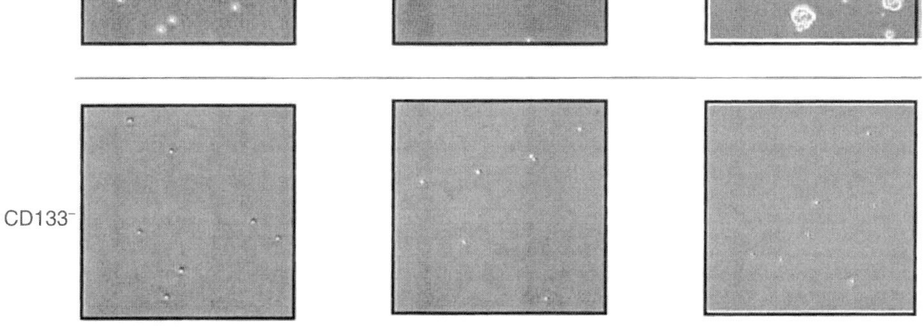

Figure 5. Flow cytometric isolation of NS-IC from fetal brain $CD133^+$, but not $CD133^-$, sorted cell start proliferating to form new neurospheres

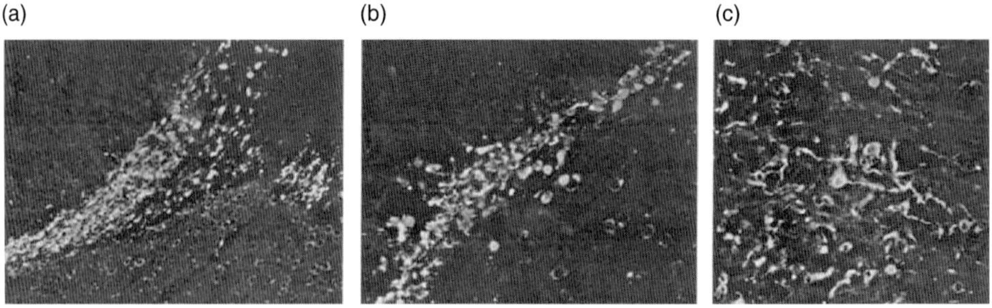

Figure 6. In vivo *studies: xenogeneic transplantation of human neural cells into NOD SCID brain 17 weeks post transplant. Progeny of* $CD133^+$ *sorted neurosphere cells reside in the subventricular zone (a), migrated through the rostral migratory stream (b) into the olfactory bulb (c). The sagittal sections were co-stained with mAbs against human nuclei with b-tubulin III.*

ventricular space of neonatal mouse brain. Seven months after transplant, human neural cells were detected with MAbs specific to human N-CAM, Thy-1 and a nuclear antigen. Human neural cells were detected in the subependymal and subventricular zones (SVZ, *Figure 6*), one of the neurogeneic sites known for proliferation of primitive progenitors in the adult brain. Human cells in the SVZ and the dentate gyrus of hippocampus co-stained with the proliferation marker Ki-67, indicating that progeny of human cells at neurogeneic sites are capable of cell division. Presence of human neural cells in the SVZ suggests that these human cells may proliferate there to continuously supply progenitor cells that migrate. In fact, human cells were detected in the rostral migratory stream and were distributed throughout the olfactory bulb (*Figure 6*). Progeny of transplanted cells were also present in the corpus callosum, cerebral cortex and in some cases, in the cerebellum. Thus, sorted/expanded human CNS-SC can engraft, proliferate, migrate and differentiate into appropriate neural lineages in response to the murine local environment. We are currently generating additional MAbs specific for human neural lineages that do not cross-react with murine brain cells. These Mabs will enable us to characterize axon formation of transplanted human neural cells, as well as identify astrocytes and oligodendrocytes in this xenogeneic transplantation system.

Prospective isolation of human NSC from fetal brain based on cell surface markers and their subsequent *ex vivo* expansion allows the potential development of a cellular replacement therapy for neurodegenerative diseases. Further identification of additional markers will define neurogenesis hierarchy in different cell populations. Analogous to the approaches successfully implemented in dissecting haemopoiesis, we would like to characterize human neural multipotent progenitors and committed progenitors for neural and glial lineages and evaluate their ability to engraft in our xenogeneic transplant model.

References

Andrews, R.G., Singer, J.W. and Bernstein, I.D. (1986) Monoclonal antibody 12-8 recognizes a 115-kd molecule present on both unipotent and multipotent hematopoietic colony-forming cells and their precursors. *Blood* **67:** 842–845.

Auerbach, J.M., Eiden, M.V. and McKay, R.D. (2000) Transplanted CNS stem cells form functional synapses in vivo. *Eur. J. Neurosci.* **12:** 1696–1704.

Baum, C.M., Weissman, I.L., Tsukamoto, A.S., Buckle, A.M. and Peault, B. (1992) Isolation of a human candidate hematopoietic stem-cell population. *Proc. Natl Acad. Sci. USA* **89:** 2804–2808.

Bertoncello, I., Hodgson, G.S. and Bradley, T.R. (1985) Multiparameter analysis of transplantable hemopoietic stem cells.I Enrichment of stem cells homing to marrow and spleen on the basis of rhodamine-123 fluorescence. *Exp Hematol* **13:** 999–1006.

Bradley, T.R. and Metcalf, D. (1966) The growth of mouse bone marrow cells in vitro. *Aust. J. Exp. Biol. Med. Sci.* **44:** 287–299.

Carpenter, M.K., Cui, X., Hu, Z.Y., Jackson, J., Sherman, S., Seiger, A. and Wahlberg, L.U. (1999) In vitro expansion of a multipotent population of human neural progenitor cells. *Exp. Neurol.* **158:** 265–278.

Civin, C.I., Strauss, L.C., Brovall, C., Fackler, M.J., Schwartz, J.F. and Shaper, J.H. (1984) Antigenic analysis of hematopoiesis. III A hematopoietic progenitor cell surface antigen defined by a monoclonal antibody raised against KG-1A cells. *J. Immunol.* **133:** 157–165.

Dexter, T.M., Allen, T.D. and Laitha, L.G. (1977) Conditions controlling the proliferation of haemopoietic stem cells in vitro. *J. Cell. Physiol.* **91:** 335–344.

Doetsch, F., Caille, I., Lim, D.A., Garcia-Verdugo, J.M. and Alvarez-Buylla, A. (1999) Subventricular zone astrocytes are neural stem cells in the adult mammalian. *Cell* **97:** 703–716.

Ezine, S., Weissman, I.L. and Rouse, R.V. (1984) Bone marrow cells give rise to distinct cell clones within the thymus. *Nature* **309:** 629–631.

Fricker, R.A., Carpenter, M.K., Winkler, C., Greco, C., Gates, M.A. and Björklund, A. (1999) Site-specific migration and neuronal differentiation of human neural progenitor cells after transplantation in the adult rat brain. *J. Neurosci.* **19:** 5990–6005.

Gage, F.H., Coates, P.W., Palmer, T.D., Kuhn, H.G., Fisher, L.J., Suhonen, J.O., Peterson, D.A., Suhr, S.T. and Ray, J. (1995) Survival and differentiation of adult neuronal progenitor cells transplanted to the adult brain. *Proc. Natl Acad. Sci. USA* **92:** 11879–11883.

Ikuta, K. and Weissman, I.L. (1992) Evidence that hematopoietic stem cells express mouse c-kit but do not depend on steel factor for their generation. *Proc. Natl Acad. Sci. USA* **89:** 1502–1506.

Johansson, C.B., Momma, S., Clarke, D.L., Risling, M., Lendahl, U. and Frisen, J. (1999) *Cell* **96:** 25–34.

Jones, R.J., Wagner, J.E., Celano, P., Zicha, M.S. and Sharkis, S.J. (1990) Identification of a neural stem cell in the adult mammalian central nervous system. *Nature* **347:** 188–189.

Kamel-Reid, S. and Dick, J.E. (1988) Engraftment of immune-deficient mice with human hematopoietic stem cells. *Science* **242:** 1706–1709.

Larochelle, A., Vormoor, J., Hanenberg, H. ***et al.*** (1996) Identification of primitive human hematopoietic cells capable of repopulating NOD/SCID mouse bone marrow: implications for gene therapy. *Nat. Med.* **2:** 1329–1337.

Lendahl, U., Zimmerman, L.B. and McKay, R.D.G. (1990) CNS stem cells express a new class of intermediate filament protein. *Cell* **60:** 585–595.

McCulloch, E.A. and Till, J.E. (1963) Repression of colony-forming ability of C57BL hematopoietic cells transplanted into non-isologous hosts. *J. Cell Comp. Physiol.* **61:** 301–308.

McCune, J.M., Namikawa, R., Kaneshima, H., Shultz, L.D., Lieberman, M. and Weissman, I.L. (1988) The SCID-hu mouse: murine model for the analysis of human hematolytic differentiation and function. *Science* **241:** 1632–1639.

Michallet, M., Philip, T., Philip, I. ***et al.*** (2000) Transplantation with selected autologous peripheral blood CD34+Thy1+ hematopoietic stem cells (HSCs) in multiple myeloma: impact of HSC dose on engraftment, safety, and immune reconstitution. *Exp. Hematol.* **28:** 858–870.

Miraglia, S., Godfrey, W., Yin, A.H., Atkins, K., Warnke, R., Holden, J.T., Bray, R.A., Waller, E.K. and Buck, D.W. (1997) A novel five-transmembrane hematopoietic stem cell antigen: isolation, characterization, and molecular cloning. *Blood* **90:** 5013–5021.

Morrison, S.J., White, P.M., Zock, C. and Anderson, D.J. (1999) Prospective identification, isolation by flow cytometry and in-vivo self-renewal of multipotent mammalian neural crest stem cells [In Process Citation]. *Cell* **96:** 737–749.

Müller-Sieburg, C.E., Whitlock, C.A. and Weissman, I.L. (1986) Isolation of two early B lymphocyte progenitors from mouse marrow: a pre-pre-B cell and a clonic Thy-1-lo hematopoietic stem cell. *Cell* **44:** 653–662.

Negrin, R. S., Atkinson, K., Leemhuis, T. *et al.* (2000) Transplantation of highly purified CD34+Thy-1+hematopoietic stem cells in patients with metastatic breast cancer. *Biol. Blood Marrow Transplant.* **6:** 262–271.

Nolta, J.A., Dao, M.A., Wells, S., Smogorzewska, E.M. and Kohn, D.B. (1996) Transduction of pluripotent human hematopoietic stem cells demonstrated by clonal analysis after engraftment in immune deficient mice. *Proc. Natl Acad. Sci. USA* **93:** 2414–2419.

Orlic, D., Anderson, S. and Bodine, D.M. (1994) Biological properties of subpopulations of pluripotent hematopoietic stem cells enriched by elutriation and flow cytometrey. *Blood Cells* **20:** 107–117; discussion 118–120.

Ploemacher, R.E., van der Sluijs, J.P., van Beurden, C.A.J., Bart, M.R.M. and Chan, P.L. (1991) Use of limiting-dilution type long-term marrow cultures in frequency analysis of marrow-repopulating and spleen colony-forming hematopoietic stem cells in the mouse. *Blood* **78:** 2527–2533.

Pluznik, D.H. and Sachs, L. (1965) The cloning of normal mast cells in tissue culture. *J. Cell Physiol.* **66:** 319–324.

Ray, J., Peterson, D.A., Schinstine, M. and Gage, F.H. (1993) Proliferation, differentiation , and long-term culture of primary hippocampal neurons. *Proc. Natl Acad. Sci. USA* **90:** 3602–3606.

Reynolds, B.A. and Weiss, S. (1992) *Science* **255:** 1707–1710.

Siminovitch, L., McCulloch, E.A. and Till, J.E. (1963) *J. Cell Comp. Physiol.* **62:** 327–336.

Sotherland, H.J., Eaves, C.J., Lansdorp, P.M., Henkelman, D.H. and Eaves, A.C. (1990) Functional characterization of individual human hematopoietic stem cells cultured at limiting dilution on supportive marrow stromal layers. *Proc. Natl Acad. Sci. USA* **87.**

Spangrude, G.J., Heimfeld, S. and Weissman, I.L. (1988) Purification and characterization of mouse hematopoietic stem cells. *Science* **241:** 58–62.

Till, J. and McCulloch, E. (1961) *Radiat. Res.* **14:** 1419–1430.

Uchida, N., Combs, J., Chen, S., Zanjani, E., Hoffman, R. and Tsukamoto, A. (1996) Primitive human hematopoietic cells displaying differential efflux of the rhodamine 123 dye have distinct biological activities. *Blood* **88:** 1297–1305.

Visser, J. W. M. and Van Bekkum, D. W. (1990) *Exp Hematol* **18:** 248–256.

Visser, J.W., Bauman, J.G., Mulder, A.H., Eliason, J.F. and de Leeuw, A.M. (1984) Isolation of murine pluripotent hemopoietic stem cells. *J Exp Med* **159:** 1576–1590.

Weilbaecher, K., Weissman, I., Blume, K. and Heimfeld, S. (1991) Culture of phenotypically defined hematopoietic stem cells and other progenitors at limiting dilution on Dexter monolayers. *Blood* **78:** 945–952.

Weissman, I.L. and Uchida, N. (2000) In: *Clinical Bone Marrow and Blood Stem Cell Transplantation* (ed. K. Atkinson). Cambridge University Press, Cambridge, pp. 19–41.

Wu, A., Till, J., Siminovitch, L. and McCulloch, E. (1967) A cytological study of the capacity for differentiation of normal hemopoietic colony forming cells. *J. Cell Physiol.* **69:** 177–184.

Wu, A.M., Till, J.E., Siminovitch, L. and McCulloch, E.A. (1968) Cytological evidence for a relationship between normal hemopoietic cell forming cells and cells of the lymphoid system. *J. Exp. Med.* **127:** 455–464.

Yandava, B.D., Billinghurst, L.L. and Snyder, E.Y. (1999) Global cell replacement is feasible via neural stem cell transplantation from the dysmyelinated shiverer mouse brain. *Proc. Natl Acad. Sci. USA* **96:** 7029–7034.

Yin, A.H., Miraglia, S., Zanjani, E.D., Almeida-Porada, G., Ogawa, M., Leary, A.G., Olweus, J., Kearney, J. and Buck, D.W. (1997) AC133, a novel marker for human hematopoietic stem and progenitor cells. *Blood* **90:** 5002–5012.

Zanjani, E.D., Pallavicini, M.G., Ascensao, J.L. *et al.* (1992) Engraftment and long term expression of human fetal hemopoietic stem cells in sheep following transplantation in utero. *J. Clin. Invest.* **89:** 1178–1188.

3

Lineage selection and isolation of neural precursors from embryonic stem cells

M. Li, D. Price and A. Smith

1. Introduction

Mouse embryonic stem (ES) cells are non-transformed immortal cell lines derived from the inner cell mass (ICM) of the preimplantation embryo (Evans and Kaufman, 1981; Martin, 1981). If grown in culture in the presence of leukaemia inhibitory factor (LIF) to prevent differentiation (Smith *et al.*, 1988), ES cells can retain their pluripotency indefinitely. This is demonstrated by their complete incorporation into the ICM and the subsequent generation of mice that are chimaeric both in somatic and germ tissues following transfer into the blastocoelic cavity of a normal blastocyst (Bradley *et al.*, 1984). The ES cell genome can be manipulated relatively at ease in culture either via the addition of new genetic information (transgene expression or gain-of-function) or via the alteration of host gene sequences (gene targeting/homologous recombination). Crucially, genetic manipulation does not compromise ES cell pluripotency and therefore mice with desired genetic modifications can be generated (Thomas and Capecchi, 1987). ES cell-based transgenic technology has greatly advanced our understanding of gene function in mammalian development.

Embryonic developments is, however, complicated and inaccessible *in utero*, and this has stimulated the exploitation of ES cells as a cellular system to study early events during development. *In vitro*, ES cells can differentiate into a wide variety of cell types during embryoid body (EB) formation, including neurons and glia (Bain *et al.*, 1995; Doetschman *et al.*, 1985; Fraichard *et al.*, 1995; Li *et al.*, 1998; Okabe *et al.*, 1996; Strubing *et al.*, 1995). It appears that the differentiating EBs follow a similar gene expression pattern to that of normal early embryo development (Bielinska *et al.*, 1996; Maltsev *et al.*, 1994). During the course of ES cell differentiation, neural precursors that express nestin and/or Sox1 and Sox2 appear first; these are followed by β-tubulin 3 and neurofilament-expressing neurons and, subsequently, GFAP or O4 expressing glial

Brain Stem Cells, edited by J. Miyan, M. Thorndyke, P.W. Beesley and C. Bannister.

cells. These results suggest that the pathway from ES cells to differentiated neural cells recapitulates, at least partially, the normal process of neurogenesis. ES cell-derived neural systems can therefore be used for experimental dissection of various aspects of mammalian neural development. If extended to humans, *in vitro* generated neural cells could also be used as a source for transplantation-based cell therapy.

However, the process of ES cell differentiation cannot be directed into a single lineage at present and non-neural cells always remain in the differentiated cultures. This presents significant problems. First, inductive or trophic signals may be masked or suppressed in such a complex cellular environment, therefore limiting the use of this system for defining the controls of neural cell commitment. Second, the presence of inappropriate cells may compromise the differentiation of the desired neurochemical phenotype in grafts following transplantation; and most undesirable is the potential for any undifferentiated ES cells present to continue proliferating and produce teratomas or even malignant teratocarcinomas (Brustle *et al.*, 1997). A possible solution to this problem would be to pre-select the population of interest in order to eliminate irrelevant cells. Below, we describe the method developed in our laboratory for the isolation of purified neural progenitors (Li *et al.*, 1998) and discuss the current status of ES cell neural differentiation.

2. From ES cells to neurons *in vitro*

2.1 *Differentiation protocols*

Different protocols have been established for the generation of neuronal and glial cells (Bain *et al.*,1995; Fraichard *et al.*, 1995; Li *et al.*, 1998; Okabe *et al.*, 1996; Strubing *et al.*, 1995) with several ES cell lines including CCE, CGR8, E14TG2a, J1, CJ7 and R1 (Doetschman *et al.*, 1985; Hooper *et al.*, 1987; Lee *et al.*, 1992; Mountford *et al.*, 1994; Nagy *et al.*, 1993; Swiatek and Gridley, 1993). Most protocols begin by culturing ES cells in medium without LIF as small embryo-like aggregates, so-called embryoid bodies (EBs). This is achieved by plating ES cells either in 'hanging drops' or in suspension in bacteriological-grade culture dishes (*Figure 1;* Doetschman *et al.*, 1985; Rudnicki and McBurney, 1987). Within these EBs, cellular derivatives of endodermal, ectodermal and mesodermal origin are differentiated. If the cultures are exposed to appropriate concentrations of retinoic acid (RA), one can induce more efficient neural differentiation (Bain *et al.*, 1995; Fraichard *et al.*, 1995; Strubing *et al.*, 1995).

The protocol used in our laboratory is derived from that of Bain *et al.* (1995; Li *et al.* (1998). ES cells were cultured as EBs in suspension in the absence of LIF for 4 days. Culture was then continued for another 4 days in the presence of 10^{-6} M RA. EBs were plated in DMEM/F12 basal medium supplemented with N2 on substrata coated with poly-D-lysine and laminin. Within 3–4 days, cells with neuronal morphology appeared. We found that prior dissociation of the EBs into single cells increase the final proportion of neural cells in the cultures was increased. Up to 70% of cells in culture were neurons even without genetic selection (described below). Comparable results were obtained with four different ES cell lines and several of their daughter lines.

A study by Strübing *et al.* (1995) used the 'hanging drop' culture method. Cells were cultivated in DMEM-DCC differentiation medium in the presence of 10^{-7} M RA during the first two days of EB culture. This was followed by 2 days of suspension culture in the absence of RA. EBs were then plated directly on an adhesive substrate.

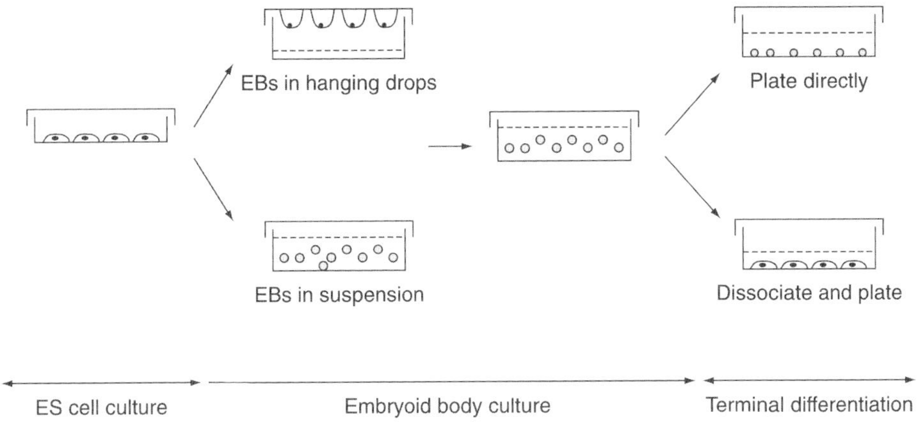

Figure 1. Differentiation of ES cells in embryoid bodies (EBs). EBs are cultured either in 'hanging drops' or in suspension. (1) Hanging drops: 10–20 µl drops of medium containing 100–1000 ES cells are placed on the lid of a culture dish. EBs form at the bottom of the drops. (2) Suspension culture: ES cells are plated in non-adhesive surface (e.g. bacterial grade dishes). After several days of aggregation culture, EBs are exposed to retinoic acid for 2–4 days. For terminal differentiation, EBs are plated on substrate either directly or following dissociation by trypsinization.

Neurons appeared about 2 days after plating. In control experiments where cells were not exposed to RA, only 15–30% of EBs contain neurons. In contrast, after induction of differentiation with RA, neurons were observed in nearly 100% of the EBs. Not all cells in the EBs, however, were neurons.

An approach to obtaining neurons from ES cells that does not use RA has also been described (Okabe *et al.*, 1996). ES cells were cultured as EBs for 4 days in standard medium in the absence of LIF, then transferred to poly-D-lysine and laminin-coated surfaces in the same medium. Following attachment of cells to the substrate, serum-containing medium was replaced with chemically defined medium. Many cells died over the next few days in culture. The surviving cultures, however, were enriched with tightly packed epithelial cells that expressed nestin. These cells proliferated in the presence of FGF2 and differentiated into both neurons and glia following its withdrawal.

2.2 *Properties of ES cell-derived neurons*

Central to the *in vitro* differentiation system is whether ES cell derivatives are phenotypically and functionally similar to their normal counterparts *in vivo*. Neurons derived from *in vitro* differentiation of ES cells were characterized by the expression of neuronal-specific markers. These include the low (NFL), middle (NFM) and the heavy (NFH) molecular mass neurofilament proteins, neural tubulin, neural transmitter genes and the synaptic vesicle protein synapsin and synaptophysin. *In vitro* generated neurons also express MAP2, Tau and GAP43, implying that they establish axons and dendrites (Bain *et al.*, 1995; Fraichard *et al.*, 1995; Finley *et al.*, 1996).

ES cell-derived neurons also exhibit complex electrophysiological properties typical of postmitotic nerve cells when examined using patch clamp (Bain *et al.*, 1995;

Fraichard *et al.*, 1995; Okabe *et al.*, 1996; Strubing *et al.*, 1995). Voltage-gated ion channels (Ca^{2+}, Na^{+} and K^{+} channels) were expressed at an early stage of ES cell-derived neurogenesis (Rohwedel *et al.*, 1998), while receptor-operated ion channels were expressed in relatively mature cultures. The expression levels of both voltage- and ligand-gated currents, measured as current density per unit capacitance, increase with time as the cells mature (Bain *et al.*, 1995; Strubing *et al.*, 1995) and are comparable to the density of functional ion channels observed in primary cultures of dissociated neural tissues (Wu and Barrish, 1994).

2.3 *ES cell-generated glia*

Many studies demonstrated the presence of glial cells following ES cell differentiation, although gliogenesis *in vitro* has not been investigated as thoroughly as neurogenesis. Most of the cultures reported so far are heterogeneous and contain neurons and other cell types, often flat cells underlying the neurons. Some of these cells are glia as shown by the expression of the astrocyte-specific marker glial fibrillary acidic protein (GFAP), or the oligodendrocyte markers GalC and O4 (Bain *et al.*, 1995. Fraichard *et al.*, 1995).

We observed that if plated in medium without mitogen, *Sox2* selected cells (to be discussed later) differentiate primarily into neurons (Li *et al.*, 1998). In cultures supplemented with serum, however, the number of cells expressing GFAP increases significantly and O4 and GalC immunopositive cells also appear. After expansion with FGF2, glial cells also tend to predominate amongst differentiated progeny (Brustle *et al.*, 1999; Li *et al.*, unpublished observation). In normal development, neuroepithelial stem cells give rise first to neurons and subsequently to glia. Our data therefore suggest that neurogenesis from ES cells *in vitro* parallels that of the normal nervous system. The appearance of neurons and glia in a temporal sequence also strongly suggests that similar multipotent precursors are generated in the Sox2 expressing population from ES cells in the course of *in vitro* differentiation.

3. Lineage selection for purified neural precursors

A major limitation with all current differentiation protocols is that the differentiation of ES cells cannot be restricted to a single cell type or even lineage, due to the disorganized and heterogeneous nature of ES cell differentiation *in vitro*. Although our protocol can generate neurons as a major population in the culture, non-neural cell types are always present. If the chemically defined medium is supplemented with serum or with basic fibroblast growth factor (bFGF or FGF2), these non-neural cells expand rapidly and progressively become the predominant cell types. Such heterogeneity complicates functional studies and molecular characterization and effectively limits the ability to propagate neural precursor cells.

To circumvent this problem, we have developed a method to purify neural precursors from mixed populations of ES cell-differentiated progeny (Li *et al.*, 1998). Since neural precursors do not express specific surface markers, as do cells of the haematopoietic system, we adopted a genetic approach termed 'lineage selection'. Although it is described here as a specific application for neural cells, lineage selection is, in principle, applicable as a general approach for isolation of other cell lineages.

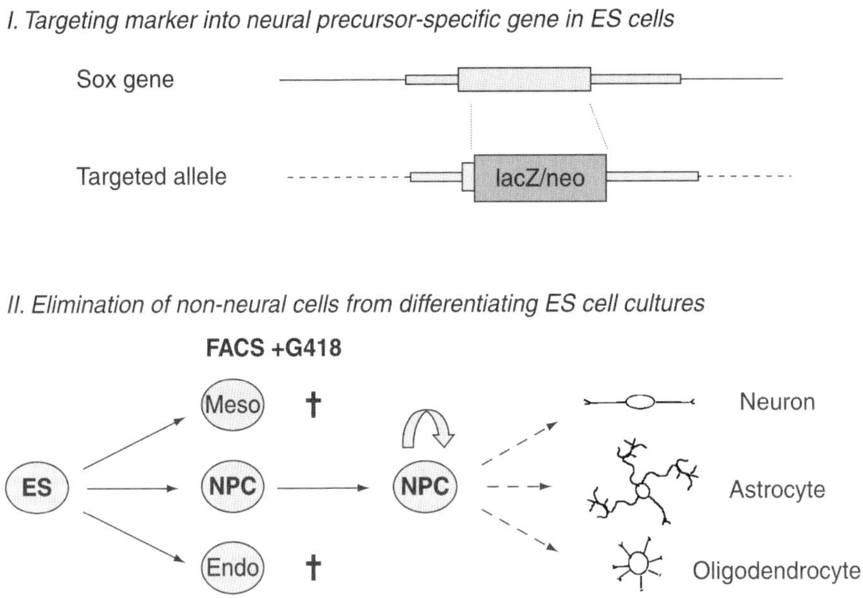

Figure 2. *Schematic representation of the lineage selection strategy for isolating pure populations of neural precursors from ES cells. Step 1: a selection marker is integrated by homologous recombination into a gene, such as certain members of the Sox gene family, that is expressed exclusively in neuroepithelium. The marker may be designed to enable fluorescence-activated cell sorting of expressing cells, for example β-galatosidase or GFP, in addition to drug selection. Step 2: following induction of ES cell differentiation, neural-restricted precursors can be isolated by cell sorting or drug selection as appropriate.*

The strategy is based on integration of a selection marker, βgeo, into the gene for the SRY-related transcription factor *Sox2* via homologous recombination (*Figure 2*). *Sox2* is a member of the family of HMG DNA binding domain-containing transcription factors related to the testis-determining gene *Sry* (Collignon *et al.*, 1996; Pevny and Lovell-Badge, 1997). Expression of *Sox2* is largely restricted to neuroepithelium. Following implantation, *Sox2* is expressed in the epiblast and extraembryonic ectoderm of the egg cylinder, becoming restricted to the prospective neural plate and chorion at the onset of gastrulation. During early somitogenesis, *Sox2* expression is confined to the entire anterior-posterior axis of the neuroectoderm and prospective sensory placodes (Collignon *et al.*, 1996; Pevny and Lovell-Badge, 1997; Wood and Episkopou, 1999). To date, *Sox2*, as well as its close relatives, *Sox1* and *Sox3*, are the most specific neuroepithelial markers identified. We have therefore chosen initially *Sox2* as a marker for the lineage selection method. A promoterless βgeo (a *LacZ-neo* fusion gene) marker was integrated into the *Sox2* locus by homologous recombination (Nicolis *et al.*, unpublished), so that only cells expressing *Sox2* (neural precursors) are βgal positive and G418 resistant. Indeed, when G418 was added to the EBs during the course of differentiation of the targeted cells, over 90% of the cells that survived showed prominent β-galactosidase reporter staining. These cells were confirmed as neuroepithelial equivalents by immunostaining for Sox1, Sox2 and nestin (*Figure 3* and *Table 1*).

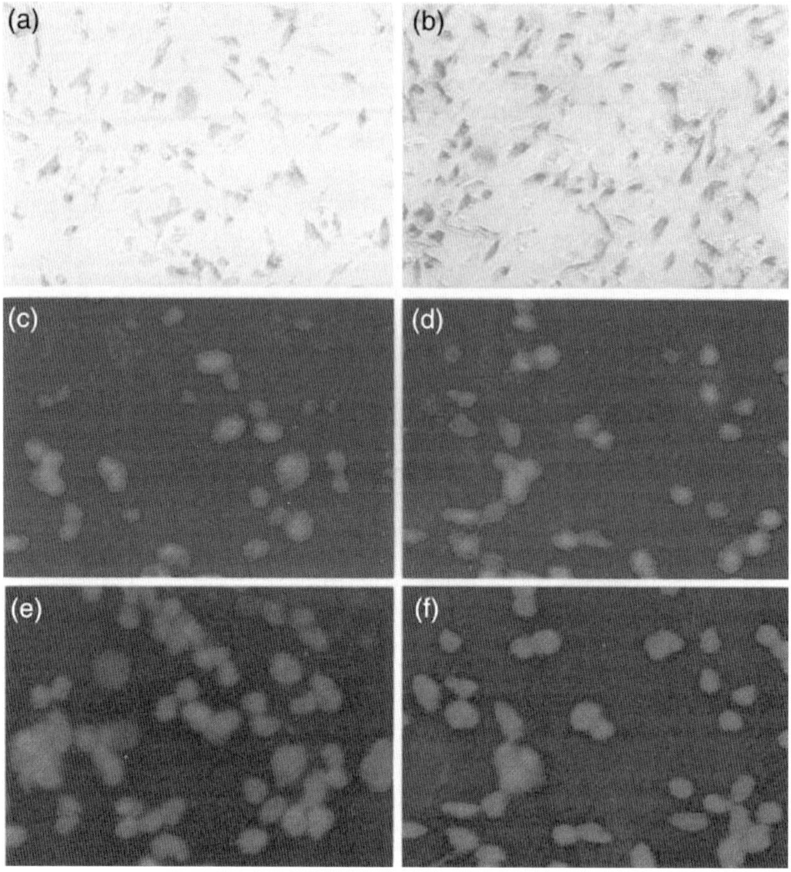

Figure 3. *Enrichment for Sox2-positive cells by G418 selection. ES cells with a targeted insertion of βgeo into Sox2 gene were induced to form embryoid bodies, and treated with retinoic acid. Cultures were (a, c, e) unselected or (b, d, f) exposed to G418 for the final 48 h of aggregate culture. Cells were analysed 3 h after plating. (a,b) Sox2-linked β-galactosidase expression in (a) an unselected culture and (b) a G418-selected culture. (c,d) Immunostaining with anti-Sox1 antibody in (c) an unselected culture and (d) a G418-selected culture. (e, f) Counterstaining with DAPI of the cultures shown in panels c and d, respectively.*

4. Properties of Sox2-selected neural precursors

4.1 *Expression of region-specific markers*

Neurogenesis is characterized by the temporally and spatially regulated expression of transcription factors. For example, the paired-box transcription factors Pax3 and Pax6 are found in dividing neural precursors throughout the length of the embryonic neural tube. Pax3 is initially expressed in the neural plate and subsequently becomes confined to the dorsal half of the neural tube (Liem *et al.*, 1995). *Pax6* is expressed from embryonic day 8 (E8) to adulthood in selected regions of the CNS, including the proliferative zone of the dorsal telencephalon (future cerebral cortex), specific parts of the ventral telencephalon, diencephalon and hindbrain, and ventrally in the spinal cord (Ericson *et al.*, 1997; Walther and Gruss, 1991; Warren and Price, 1997). Widespread expression of both

Table 1. Expression of neural markers in Sox2*-selected cultures*

Marker	Percentage of positive cells
Sox2	95.5±1.1 (510)
Sox1	88.7±5.1 (490)
Nestin	94.0±3.2 (298)
Pax6	46.2±9.8 (764)
Pax3	35.5±8.8 (319)
Mash-1	24.2±7.6 (361)
Math-1	5.1±1.45 (316)
Math-4A	13.8±6.3 (679)
Math 3C	9±2.9 (444)
Delta-1	30.6±4.7 (72)
NeuroD	4±1.6 (450)
Islet-1	2.8±0.9 (901)

Cells incubated in the presence of G418 for the final 48 h of embryoid body culture were analysed 3–4 h after plating, except for Islet 1 detection which was performed after 24 h. Data for Sox1, Sox2, Sox3, Pax6 and Mash1 are averaged from two independent experiments, and for the other markers are from duplicate wells from a single experiment. The numbers within parenthesis indicate the number of cells scored.

Pax genes was found in *Sox2*-selected cells analysed on the day of plating (*Table 1*). This suggests that ES cell-derived neural precursors can acquire both dorsal and ventral identities. In the developing brain, it is thought that the presence of the bHLH factors Mash1 and math-4A defines distinct subpopulations of neural progenitors, because the distribution of these two transcription factors is mutually exclusive (Gradwohl *et al.*, 1996). Expression of both genes was detected in sub-populations of *Sox2*-selected cultures (*Table 1*). Other regulatory genes expressed during normal neurogenesis were also expressed in populations of *Sox2*-selected cultures, including *math-1, math-4a, math-4c, Notch1* and its ligand *Delta I* (*Table 1;* Li *et al.*, unpublished). These results suggest that the diversity of neural differentiation could be achieved from ES cells in the absence of embryonic axial organization and that the *Sox2*-selected cultures appear to reflect the range of sub-specification of progenitors within the neural tube (Tanabe and Jessell, 1996). Requirements for specification and differentiation of individual progenitor types could therefore be investigated in this system, both by addition of extracellular factors and by genetic manipulation of the ES cells prior to differentiation. These results also suggest that lineage selection strategy could be adapted to isolate stem or precursor cells for other lineages. An important attribute is that unlike immunopurification techniques this approach is not limited to cell-surface antigens but could be applied to any differentially expressed gene regardless of the nature of its product. The most obvious potential is for exploiting the growing number of genes that have been identified that encode transcription factors with lineage-restricted expression. Selected populations could readily be refined by introducing independent markers into more than one gene.

4.2 *Proliferation and differentiation capacity* in vitro

The co-expression of *Sox* and *Pax* genes is associated with the proliferative neuroectoderm in developing embryos. FGF2 has been shown to have a mitogenic effect on dissociated primary neural precursors *in vitro* (Antonopoulos *et al.*, 1997; Cavanagh *et al.*, 1997; Ghosh and Greenberg, 1995; Kilpatrick and Bartlett, 1993), and injecting FGF2 into the cerebral ventricles of rat embryos increases the number of rounds of division of cortical progenitors (Vaccarino *et al.*, 1999). When cultured in medium supplemented with FGF2, *Sox2*-selected cells divide, retain undifferentiated morphology and show strong X-gal staining, indicative of *Sox2* expression in predominant populations (*Figure 4*). Such cultures could be amplified and serially passaged for at least 4 weeks.

(a)
(b)
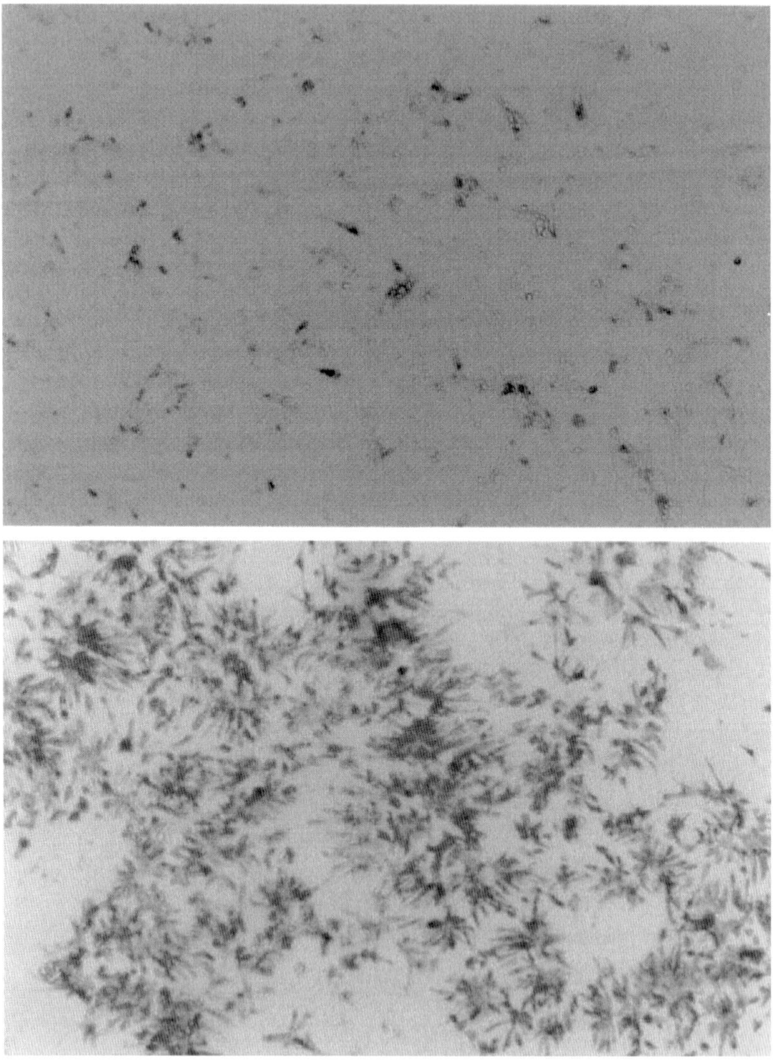

Figure 4. *Sox2-positive cells proliferate* in vitro *in response to bFGF. Following 8 days induction of neural differentiation, cells were dissociated by trypsinization, plated at low density, and cultured in DMEM/F12 supplemented with N2 medium in the presence of 10 ng ml*$^{-1}$ *of bFGF and 200 mg ml*$^{-1}$ *of G418. (a) Overnight culture; (b), 4-day culture. Note that the cell numbers increased around 10-fold over the culture period, almost all of the cells expanded by bFGF maintained Sox2 expression as shown via an X-gal stain.*

In the absence of mitogen (FGF2), *Sox2*-selected cells differentiate and give rise to all types of neural cells (i.e. neurons, astrocytes and oligodendrocytes). The early neuronal marker β-tubulin III can be detected 1–2 days after plating. In 4-day culture, most neuronal-like cells express neurofilament light chain protein (NFL) and MAP2, markers associated with functional nerve cells such as synaptic proteins (synapsin I and synaptophysin). The neurotransmitters γ-amino butyric acid (GABA) and glutamate were observed in more mature cultures (1–2 weeks after plating).

5. Plasticity of ES cell derivatives

One way of assessing the competence of *Sox2*-selected neural precursors to generate appropriate types of neural cell is to transplant them into the developing brain, where they will be exposed to a wide range of signals. In addition, this approach takes work on ES-derived neural precursors closer to an application in cell therapy for developmental or neurodegenerative disorders. This is a long-term aim based on the notion that self-renewing, pluripotent ES cells may provide an unlimited donor source for transplantation (Smith, 1998).

In recent years, several groups have developed methods for injecting cells into the telencephalic vesicles of embryonic rodents (Brustle *et al.*, 1995; Campbell *et al.*, 1995; Caric *et al.*, 1997; Cattaneo *et al.*, 1994). For example, to study the regulation of neuronal positioning, Brustle *et al.* (1995) injected primary neural precursors derived from dorsal and ventral mouse forebrain into the telencephalic vesicles of embryonic rats. Cells from both areas participated in the generation of telencephalic, diencephalic, and mesencephalic brain regions. Donor-derived neurons populated the host brain and acquired phenotypic features appropriate for their final location, indicating that neuronal migration and differentiation are predominantly regulated by non-cell-autonomous signals. Such work established that intrauterine transplantation by injection of primary cells into the cerebral vesicles allows the generation of chimaerism in the mammalian brain. With the use of ultrasound guidance, targeted intraparenchymal injections into even very young mouse embryos is now possible, and has been used to test the degree of commitment present in primary neural progenitors from the embryonic brain (Olsson *et al.*, 1997).

In 1997, Brustle *et al.* reported the use of these methods to transplant ES cells that had been differentiated into neural precursors in culture into the ventricles of embryonic rats, as a means of testing their developmental potential. As with primary neural precursors, over the weeks following injection the transplanted ES cell-derived cells migrated in large numbers into brain tissue and gave rise to neurons, astrocytes and oligodendrocytes. They incorporated into telencephalic, diencephalic, and mesencephalic regions and assumed phenotypes indistinguishable from neighbouring host cells. The transplanted cells remaining in the host ventricle formed large clusters containing prominent intraventricular neuroepithelial structures. These formations closely resembled tightly packed neural tubes and individual tubes consisted of columnar epithelium with high mitotic activity at the luminal surface. The epithelial cells showed strong expression of antigens typically expressed in the early neuroepithelium, such as nestin. Cells expressing neuronal antigens, including microtubule-associated protein 2, neurofilament, and tyrosine hydroxylase were restricted to the periphery of the formations, suggesting an inside-out gradient of differentiation, as occurs in the developing neuroepithelium *in vivo*. Brustle *et al.* (1997) noted that their intraventricular grafts contained small clusters of still undifferentiated embryonic cells that decreased in number with age, while there was a gradual emergence of non-neural tissue with age. Occasional islands containing adenoid structures, cartilage and epidermis were found within the intraventricular clusters, indicating that undifferentiated embryonic cells present in the transplanted cell suspension form intraventricular teratomas.

This work demonstrated that *in vitro*-generated ES cell-derived neural precursors are able to respond to environmental signals guiding cell migration and differentiation in the intact brain, where they have the potential to reconstitute neuronal and glial lineages in the central nervous system. Subsequent work has led to the generation of ES cell-derived precursors for astrocytes and oligodendrocytes that can efficiently myelinate axons in brain and spinal cord in a rat model of a human myelin disease (Brustle *et al.*, 1999). While all of this is hugely encouraging in the search for methods of cell therapy for human disease, significant problems remain. One of these is that previous transplants of ES-derived neural precursors have led to the generation of intraventricular teratomas in the forebrain, an undesirable outcome in any therapeutic application. This may be avoided if the ES-derived neural precursors are selected from the non-differentiated ES cells prior to transplant, as in the *Sox2*-selection procedure described above.

We have carried out transplants of *Sox2*-selected ES cell-derived neural precursors into the telencephalic vesicles of E16 rats. The host animals were born and survived for a week after birth, at which point their brains were processed for histology. The transplanted cells were labelled either with a fluorescent dye or with bromodeoxyuridine or expressing lacZ and/or GFP constitutively. We found transplanted cells integrated into the brain tissue and showing a neuronal appearance when fluorescently labelled. We also observed the generation of neural tube-like structures in the ventricular spaces. These had appearances similar to those described for non-selected ES cell-derived precursors by Brustle *et al.* (1999), but lacked any obvious evidence of non-neural tissue (*Figure 5*).

Our conclusion is that the *Sox2*-selected ES cell-derived neural precursors can respond to the host environment in such a way as to allow their integration into brain tissue and the formation of neuroepithelial structures. Our hope is that these cells will not generate intraventricular teratomas and ongoing work is aimed at demonstrating this.

6. Future prospects

The ES cell *in vitro* technology allows the study of cellular differentiation processes during neurogenesis in culture, from the founder stem cells of the embryo (ES cells), via lineage restricted stem cells (neural stem cells) into highly differentiated and specialized neurons. This system combines several strengths: reliable and sophisticated genetic manipulation systems are available for ES cells; there is the opportunity to obtain large numbers of cells, permitting biochemical and molecular analysis. In addition, there is the future prospect of using human ES cell derivatives for therapeutic transplantation. The derivation of multipotent neural precursors from ES cells is therefore of great interest. The ES cell-derived neural system is made even more attractive by the ability to generate pure populations of neural precursors and thereafter highly enriched neurons by means of lineage selection.

During the last decade, many genes that may play important roles in nervous system development and function have been knocked out. As an alternative and/or

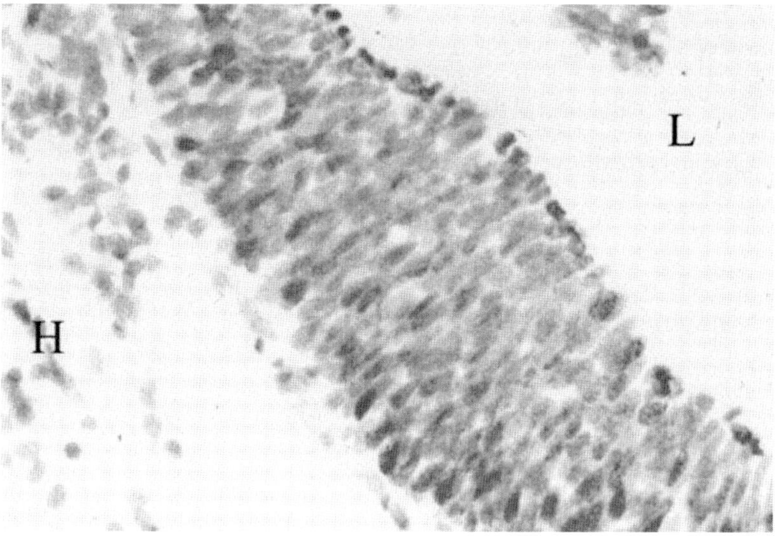

Figure 5. *Neuroepithelial structures in the ventricular space of early postnatal rats transplanted with Sox2-selected ES derivatives at E16. Section through the forebrain of a postnatal day 5 rat whose telencephalic vesicle was injected with ES cell-derived neural precursors on E16. The injected cells formed structures characteristic of the developing neural tube. Mitotic figures are evident along the luminal (L) edge and cells in the wall show radial elongation typical of migrating neurons. H, host cells.*

complementary approach to making mice, neural differentiation of ES cells *in vitro* offers an attractive way to analyse the consequences of mutations during early neurogenesis. Indeed EBs may be an invaluable approach to assess neural function if the targeted ablation results in lethality in development. 'Gain of function', on the other hand, could be used to express possible regulatory and determination genes in ES cells followed by examination of the consequences for neural induction and differentiation. One such example is the expression of neuroD2 and neuroD3, which has been reported to turn ES cells into neurons without RA treatment (O'Shea *et al.*, 1997). For pharmaceutical considerations, one could express mutant genes found in human diseases in ES cells and use the *in vitro* culture system to screen for possible therapeutic compounds.

Effects of growth factors and inductive molecules on neural stem cell fate determination and differentiation can be analysed in the culture system described here. If combined with genetic manipulation (e.g., expression of a cDNA library, ideally in an inducible manner), ES cell *in vitro* technology could potentially be used to screen for novel genes governing neural stem cell maintenance and differentiation, or as an experimental system to study such candidate genes. *In vitro* neurogenesis can also be used as a developmental model to analyse the expression of a large number of genes during neural induction and terminal differentiation by means of differential display, or cDNA microarray expression profiling (Fodor *et al.*, 1993; Gress *et al.*, 1992; Kelly and Rizzino, 2000; Lennon and Lehrach, 1991; Schena *et al.*,1995).

ES cell derivatives can colonize host tissue and differentiate after transplantation into adult rodent recipients (Deacon *et al.*, 1998; Dinsmore *et al.*, 1996). Neural cells derived from human ES cells would therefore provide a source of cells for transplantation-based cell therapy for neurodegenerative disorders such as Parkinson's and Huntington's diseases (Reubinoff *et al.*, 2000). Recently, tyrosine hydroxylase-expressing neurons have been shown to be generated from ES cells following induction with sonic hedgehog and FGF8 (Lee *et al.*, 2000; Ye and Shimamura, 1998). These are the types of neurons lost in Parkinson's disease patients and therefore would be of special interest for therapeutic purposes. Integration of a mesencephalic dopaminergic lineage marker into ES cells prior to their differentiation may make it possible to obtain homogeneous populations of this cell type for grafting.

Acknowledgements

Research in the authors' laboratories is supported by the MRC and HFSP.

References

Antonopoulos, J., Pappas, I.S. and Parnavalas, J.G. (1997) Activation of the GABA receptor inhibits the proliferative effects of bFGF in cortical progenitor cells. *Eur. J. Neurosci.* **9**: 291–298.

Bain, G.D., Kitchens, D., Yao, M., Hunttner, J.E. and Gottlieb, D.I. (1995). Embryonic stem cells express neuronal properties *in vitro*. *Dev. Biol.* **168**: 342–357.

Bielinksa, M., Narita, H., Heikinheimo, M., Porter, S.B. and Wilson, D.B. (1996) Erythropoiesis and vasculogenesis in embryoid bodies lacking visceral yolk sac endoderm. *Blood* **88**: 3720–3730.

Bradley, A., Evans, M.J., Kaufman, M.H. and Robertson, E. (1984) Formation of germ-line chimaeras from embryo-derived teratocarcinoma cell lines. *Nature* **309**: 255–256.

Brustle, O., Maskos, U. and McKay R.D.G. (1995) Host-guided migration allows targeted introduction of neurons into the embryonic brain. *Neuron* 15: 1275–1285.

Brustle, O., Spiro, AC., Karram, K., Choudhary, K., Okabe, S. and McKay R.D.G. (1997) *In vitro*-generated neural precursors participate in mammalian brain development. *Proc. Natl Acad Sci. USA* **94**: 14809–14814.

Brustle O., Jones K.N., Learish R.D., Karram K., Choudhary K., Wiestler O.D., Duncan I.D. and McKay R.D. (1999) Embryonic stem cell-derived glial precursors: a source of myelinating transplants. *Science* **285**: 754–746.

Campbell, K., Olsson, M. and Björklund A. (1995) Regional incorporation and site-specific differentiation of striatal precursors transplanted to the embryonic forebrain ventricle. *Neuron* **15**: 1259–1273.

Caric, D., Gooday, D., Hill, R.E., McConnell, S.K. and Price, D.J. (1997) Determination of the migratory capacity of embryonic cortical cells lacking the transcription factor Pax-6. *Development* **124**: 5087–5096.

Cattaneo, E., Magrassi, L., Butti, G., Santti, L., Giavazzi, A. and Pezzotta, S. (1994) A short term analysis of the behaviour of conditionally immortalized neuronal prognitors and primary neuroepithelial cells implanted into the fetal rat brain. *Dev. Brain. Res.* **83**: 197–208.

Cavanagh, J.F.R., Mione, M.C., Pappas, I.S. and Parnavalas, J.G. (1997) Basic fibroblast growth factor prolongs the proliferation of rat cortical progenitor cells in vitro without altering their cell cycle parameters. *Cerebral Cortex* **7**: 293–302.

Collignon, J., Sochanathan, S., Hacker, A., Cohen-Tannoudji, M., Norris, D., Rastan, S., Stefanovic, M., Goodfellow, P. and Lovell-Badge, R. (1996) A comparison of the properties of *Sox3* with *Sry* and two related genes. *Sox1* and *Sox2*. *Development* 122: 509–520.

Deacon, T., Dinsmore, J., Costantini, L., Ratliff, J. and Isacson, O. (1998) Blastula-stage stem cells can differentiate into dopaminergic and serotonergic neurons after transplantation. *Exp. Neuro.* **149**: 28–41.

Dinsmore, J., Ratliff, J., Deacon, T., Pakzaban, P., Jacoby, D., Galpern, W. and Isacson, O. (1996). Embryonic stem cells differentiated *in vitro* as a novel source of cells for transplantation. *Cell Transplantation* **5:** 131–141.

Doetschman, T.C., Eistertter, H., Katz, M., Schmidt, W. and Kemler, R. (1985) The *in vitro* development of blastocyst-derived embryonic stem cell lines: formation of viserval yolk sac, blood islands and myocardium. *J. Embryol. Exp. Morphol.* **87:** 27–45.

Ericson, J.P., Rashbass, A., Schedl, S., Brenner-Morton, A., Kawakami, V., van Heyningen, T.M., Jessell, T. and Briscoe, J. (1997) Pax6 controls progenitor cell identity and neuronal fate in response to graded Shh signaling. *Cell* **90:** 169—180.

Evans, M.J. and Kaufman, M.H. (1981) Establishment in culture of pluripotential cells from mouse embryos. *Nature* **292:** 154–156.

Finley, M.F., Kulkarni, N. and Huettner, J.E. (1996) Synase formation and establishment of neuronal polarity by P19 embryonic carcinoma cells and embryonic stem cells. *J. Neurosci.* **16:** 1056–1065.

Fodor, S.P., Rava, R.P., Huang, X.C., Pease, A.C., Holmes, C.P. and Adams, C.L. (1993) Multiplexed biochemical assays with biological chips. *Nature* **364**: 555–564.

Fraichard, A., Chassande, O., Bilbaut, G., Dehay, C., Savatier, P. and Samarut, J. (1995) *In vitro* differentiation of embryonic stem cells into glial cells and functional neurons. *J. Cell Sci.* **108:** 3181–3188.

Ghosh, A. and Greenberg, M.E. (1995) Distinct roles for bFGF and NT-3 in the regulation of cortical neurogenesis. *Neuron* **15:** 89–103.

Gradwohl, G., Fode, D. and Guillemot, F. (1996) Restricted expression of a novel murine atonal-related bHLH protein in undifferentiated neural precursors. *Dev. Biol.* **180:** 227–241.

Gress, T.M., Hoheisel, J.D., Lennon, G.G., Zehetner, G. and Lehrach, H. (1992) Hybridization fingerprinting of high-density cDNA-library arrays with cDNA pools derived from whole tissues. *Mamm. Genome* **3:** 609–619.

Hooper, M., Hardy, J., Handyside, A., Hunter, S. and Monk, M. (1987) HPRT-deficient (Lesch-Nyhan) mouse embryos derived from germline colonization by cultured cells. *Nature* 326: 292–295.

Kelly, D. and Rizzino, A. (2000) DNA microarray analyses of genes regulated during the differentiation of embryonic stem cells. *Mol. Rep. Dev.* **56:** 113–123.

Kilpatrick, T.J. and Bartlett, P.F. (1993) Cloning and growth of multipotential precursors: requirements for proliferation and differentiation. *Neuron* 10: 255–265.

Lee, K.F., Li, E., Huber, J., Landis, S., Sharpe, A.H., Chao, M.V. and Jaenisch, R. (1992) Targeted mutation of the gene encoding the low affinity NGF receptor P75 leads to deficits in the peripheral sensory nervous system. *Cell* 69: 737–749.

Lee, S., Lumelsky, N., Studer, L., Auerbach, J. and McKay, R.G.D. (2000) Efficient generation of midbrain and hindbrain neurons from mouse embryonic stem cells. *Nature Biotechnol.* 18: 675–679.

Lennon, G.G. and Lehrach, H. (1991) Hybridization analysis of arrayed cDNA libraries. *Trends Genet.* 7: 314–317.

Li, M.L., Pevny, L., Lovell-Badge, R. and Smith, A. (1998).Generation of purified neural precursors from embryonic stem cells by lineage selection. *Curr. Biol.* **8**: 791–794.

Liem, K., Tremmi, G., Roelink, H. and Jessell, Y. (1995) Dorsal differentiation of neural plate cells by BMP-mediated signals from epidermal ectoderm. *Cell* **82:** 969–979.

Lundberg, C.A., Martinez-Serrrano, A., Cataneo, E., McKay, R.G.D. and Björklund, A. (1997). Survival, integration, and differentiation of neural stem cell lines after transplantation to the adult rat striatum. *Exp. Neurol.* **145:** 342–360.

Martin, G.R. (1981) Isolation of a pluripotent cell line from early mouse embryos cultured in medium conditioned by teratocarcinomal stem cells. *Proc. Natl Acad. Sci. USA* **78:** 7634–7638.

Mlatsev, V.A., Wobus, A.M., Rohwedl, J., Bader, M. and Hescheler, J. (1994) Cardiomyocytes differentiated *in vitro* from embryonic stem cells developmentally express cardiac-specific genes and ionic currents. *Circ. Res.* **75:** 233–244.

Mountford, P., Zevnik, B., Düwel, A., Nichols, J., Li, M., Dani, C., Robertson, M., Chambers, I. and Smith, A. (1994) Dicistronic targeting constructs: Reporters and modifiers of mammalian gene expression. *Proc. Natl Acad. Sci. USA* **91:** 4303–4307.

Nagy, A., Rossant, J., Nagy, R., Abramow-Newerly, W. and Roder, J.C. (1993) Derivation of completely cell-culture-derived mice from early-passage embryonic stem cells. *Proc. Natl Acad. Sci USA* **90:** 8424–8428.

Okabe, S., Fosberg-Nilsson, K., Spiro, A.C., Sega, M. and McKay, R.G.D. (1996) Development of neuronal precursors and functional postmitotic neurons from embryonic stem cells in vitro. *Mech. Dev.* **59**: 89–102.

Olsson, M., Campbell, K. and Turnbull, D.H. (1997) Specification of mouse telencephalic and mid-hindbrain progenitors following heterotopic ultrasound-guided embryonic transplantation. *Neuron* **19:** 761–772.

O'Shea, K.S., Gratsch, T.E., Tapscott, S.J. and McCormick, M.B. (1997) Neuronal differentiation of embryonic stem (ES) cells constitutively expressing neuroD2 or neuroD3. *Soc. Neurosci. Abstr.* **23:** 1144–1148.

Pevny, L. and R. Lovell-Badge (1997). Sox genes find their feet. *Curr. Opin. Genet. Dev.* 7: 338–344.

Reubinoff B.E., Pera M.F., Fong C.Y., Trounson A. and Bongso, A. (2000) Embryonic stem cell lines from human blastocysts: somatic differentiation *in vitro*. *Nat. Biotechnol.* **18:** 399–404.

Rohwedel, J., Kleppisch, T., Pich, U. *et al.* (1998) Formation of postsynaptic-like membranes during differentiation of embryonic stem cells *in vitro*. *Exp. Cell Res.* **239:** 214–225.

Rudnicki, M.A. and McBurney, M.W. (1987) Cell culture methods and induction of differentiation of embryonic carcinoma cell line. In: *Teratocarcinomas and Embryonic Stem Cells* (ed. E.J. Robertson), pp. 19–49. IRL Press, Washington DC, USA.

Schena, M., Shalon, D., Davis, R.W. and Brown, P.O. (1995) Quantitative monitoring of gene expression patterns with a complementary DNA microarray. *Science* **270:** 368–371.

Smith, A. (1998) Cell therapy: In search of pluripotency. *Curr. Biol.* **8**: 802–804.

Smith, A., Heath, J.K., Donaldson, D.D., Wong, G.G., Moreau, J., Stahl, M. and Rogers, D. (1988) Inhibition of pluripotential embryonic stem cell differentiation by purified polypeptides. *Nature* **336:** 668–690.

Strubing, C., Ahnert-Hilger, G., Jin, S., Wiedenmann, B., Hescheler, J. and Wobus, A.M. (1995) Differentiation of pluripotent embryonic stem cells into the neuronal lineage *in vitro* gives rise to mature inhibitory and excitatory neurons. *Mech. Dev.* **53:** 275–287.

Swiatek, P.J. and Gridley, T. (1993) Perinatal lethality and defects in hindbrain development in mice homozygous for a targeted mutation of the zinc finger gene *krox*20. *Genes Dev*. **7:** 2071–2084.

Tanabe, Y. and Jessell, T. (1996) Diversity and pattern in the developing spinal cord. *Science* **274:** 1115–1123.

Thomas, K.R. and Capecchi, M.R. (1987) Site-directed mutagenesis by gene targeting in mouse embryo-derived stem cells. *Cell* **51:** 503–512.

Vaccarino, F.M., Schwartz, M.L., Raballo, R., Nilsen, J., Rhee, J., Zhou, M., Doetschman, T., Coffin, J.D., Wyland, J.J. and Hung, Y.-T.E. (1999) Changes in cerebral cortex size are governed by fibroblast growth factor during embryogenesis. *Nature Neurosci.* **2:** 246–253.

Walther, C. and Gruss, P. (1991) *Pax-6*, a murine-paired box gene, is expressed in the developing CNS. *Development* **113:** 1435–1449.

Warren, N. and Price, D.J. (1997) Roles of Pax-6 in murine diencephalic development. *Development* **124:** 1573–1582.

Wood, H.B. and Episkopou, V. (1999) Comparative expression of the mouse Sox1, Sox2 and Sox3 genes from pre-gastrulation to early somite stages. *Mech. Dev.* **86:** 197–201.

Wu, R.L. and Barrish, M.E. (1994) Astroglial modulation of transient potassium current development in cultured mouse hippocampal neurons. *J. Neurosci.* **14:** 1677–1687.

Ye, W., Shimamura, K., Rubenstein, J.L. *et al.* (1998) FGF and Shh signals control dopaminergic and serotonergic cell fate in the anterior neural plate. *Cell* **93:** 755–766.

4

Early generated Cajal–Retzius cells have different functions in cortical development

M. Frotscher, L. Seress, H. Abraham and B. Heimrich

1. Introduction

Cajal–Retzius (CR) cells are early-generated neurons that play an important role in neuronal migration by secreting the glycoprotein reelin. In this article, we summarize recent data that point to additional functions of CR cells. We demonstrate that CR cells are essential for the pathfinding of entorhinal axons to the hippocampus and fascia dentata and for the recognition of the correct target layers. In both pathfinding and target recognition, however, not reelin but other, as yet unknown, signals appear to be of primary importance. We discovered an early projection of hippocampal CR cells to the entorhinal cortex and hypothesize that this pioneer projection may provide a template for later outgrowing entorhinal fibres to reach their hippocampal target areas. The formation of early transient synapses with CR cell somata and dendrites, located in the appropriate termination zones of entorhinal fibres, may be an important step in the target recognition of these axons. Reelin was shown to affect the branching of entorhinal axons and synapse formation. Taken together, these data demonstrate that CR cells and reelin serve a variety of functions in crucial steps of brain development.

When Ramón y Cajal and Gustaf Retzius described a particular class of neurons in layer 1 of the cortex (Ramón y Cajal, 1891, 1911; Retzius, 1893, 1894), they were struck by their mainly horizontal orientation which contrasts to the vertical orientation of most other neurons in the cortical mantle. The observation that many of these CR cells undergo degeneration during the early postnatal period (e.g., Derer and Derer, 1990) suggested a role for CR cells in development, but their function remained obscure. With the discovery that CR cells synthesize and secrete the glycoprotein reelin (D'Arcangelo *et al.*, 1995, 1997; Hirotsune *et al.*, 1995; Ogawa *et al.*, 1995), it became obvious that CR cells have an important function in neuronal migration. In the mouse mutant reeler lacking reelin, the regular lamination of cortical neurons is

Brain Stem Cells, edited by J. Miyan, M. Thorndyke, P.W. Beesley and C. Bannister.

severely disturbed indicating that this glycoprotein is important for the migration of cortical neurons to their destinations in distinct layers of the cerebral cortex (Frotscher, 1997, 1998; Rakic and Caviness, 1995). In addition to its role in neocortical lamination, reelin was found to be involved in migratory processes in the hippocampus and cerebellum. Recent studies summarized in this report suggest that CR cells have additional functions. They are important for pathfinding and target recognition of a distinct class of axons innervating the hippocampus and fascia dentata, the fibres from the entorhinal cortex (Ceranik *et al.*, 1999, 2000; Del Rio *et al.*, 1997). These latter findings illustrate that early pioneer neurons such as CR cells are likely to govern a variety of different developmental processes.

2. Pathfinding, target recognition and synapse formation of entorhino-hippocampal axons

Many neuroanatomists were impressed by the route that fibres from the entorhinal cortex choose to find their hippocampal target areas. They traverse the subiculum and terminate with high precision in the superficial layers of the hippocampus proper and dentate gyrus. What are the signals that guide these fibres from the entorhinal cortex to the hippocampus and allow them to recognize their appropriate target layers? One way to approach these questions is by cutting this highly complicated process into single steps. Thus, as a first step, the entorhinal fibres have to find their way to the hippocampus, which we will call *pathfinding* here. Next, they have to know where to stop, a process known as *target recognition*. Finally, the tip of the growing axon, the growth cone, will form a presynaptic bouton that establishes a synapse. This latter process accordingly comprises what we call *synapse formation*. As demonstrated below, CR cells and their glycoprotein reelin are likely to be involved in all three processes. We focused on these neurons, because CR cells of the hippocampus and dentate gyrus, like CR cells in the neocortex, are located in the marginal zone which in the hippocampal formation corresponds to the termination field of entorhinal axons. Could it be that the transient presence of CR cells in these layers is linked to the layer-specific ingrowth of entorhinal axons?

3. Not granule cells, but CR cells, are important for the target recognition of entorhinal axons

Entorhinal axons are known to invade the hippocampus during the late embryonic period (Supèr and Soriano, 1994). This early arrival of entorhinal fibres contrasts with the late formation of their proper target cells, the dentate granule cells. As known from many studies, the majority of granule cells are formed postnatally and, thus, these neurons are unlikely to assist the early-arriving entorhinal fibres in recognizing their appropriate target zone. In spite of this, the entorhinal fibres early on recognize the outer portion of the molecular layer as their termination field. To further substantiate that the granule cells are not important for target layer recognition, we x-irradiated new-born rats resulting in a severe loss of mitotically active granule cell precursors. Despite a dramatic loss of granule cells, in some cases an almost complete elimination of the granular layer, the entorhinal axons, labelled by anterograde transport of biotinylated dextran amine (BDA), as normal recognized their target zone in the dentate gyrus (*Figure 1*). We take this as evidence that the granule cells are not, in fact, involved in pathfinding and target recognition of entorhinal fibres.

What about CR cells? Unlike the granule cells, CR neurons are generated very early during embryonic development (Del Rio *et al.*, 1996; Meyer *et al.*, 1998). Thus, they are present at the time of entorhinal fibre ingrowth and could serve as transient targets. In order to test this, we used slice cultures of hippocampus, co-cultured with entorhinal cortex, as a model system. Previous studies had already shown that under these *in vitro* conditions the entorhinal fibres recognize their target layers, indicating that the relevant guiding signals are present (Frotscher and Heimrich, 1993, 1995; Li *et al.*, 1993). As a first step, we were able to show that entorhinal axons in entorhino-hippocampal co-cultures established transient synaptic contacts with CR cells in the outer molecular layer (Del Rio *et al.*, 1997; *Figure 2*). This suggested that CR cells, by their synapses with the entorhinal axons, keep these fibres in their appropriate termination zone, the outer molecular layer. In order to test this further, we attempted to eliminate CR cells. In previous studies (Del Rio *et al.*, 1996) we had observed that CR cells are highly sensitive to glutamate receptor agonists. Thus, we exposed slice cultures of hippocampus to the glutamate receptor agonist, domoic acid, in order to induce a rapid CR cell degeneration. Then, such slice cultures lacking CR cells were confronted with ingrowing entorhinal axons supplied by a co-culture of entorhinal cortex. The result was very clear: under these conditions, the entorhinal axons were

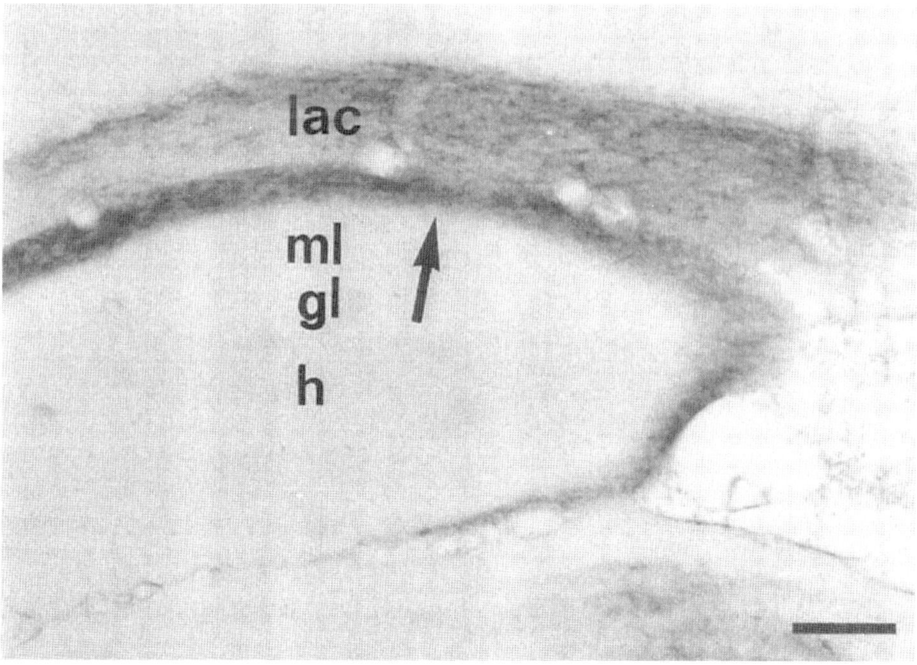

Figure 1. *Fascia dentata of a rat subjected to x-ray irradiation in the neonatal period. Despite a dramatic reduction in the number of dentate granule cells, BDA-labelled entorhinal axons found their appropriate target zones, the outer portion of the dentate molecular layer (ml, arrow) and the stratum lacunosum-moleculare (lac) of the hippocampus proper. gl, remnant of granular layer; h, hilus. Scale bar: 200 μm.*

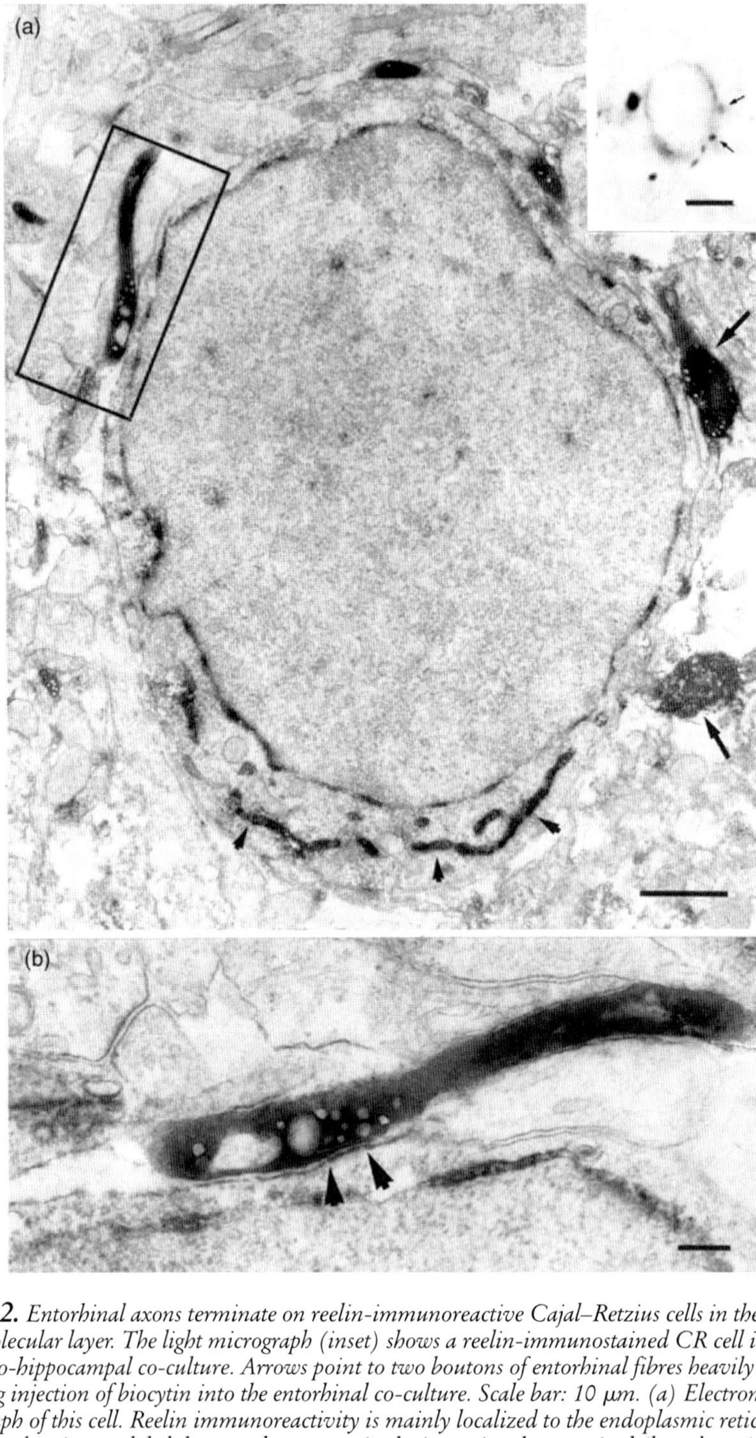

Figure 2. Entorhinal axons terminate on reelin-immunoreactive Cajal–Retzius cells in the dentate outer molecular layer. The light micrograph (inset) shows a reelin-immunostained CR cell in an entorhino-hippocampal co-culture. Arrows point to two boutons of entorhinal fibres heavily labelled following injection of biocytin into the entorhinal co-culture. Scale bar: 10 μm. (a) Electron micrograph of this cell. Reelin immunoreactivity is mainly localized to the endoplasmic reticulum (arrowheads). Arrows label the same boutons as in the inset. Another terminal (boxed area) is shown at higher magnification in (b). Scale bar: 1 μm. (b) Higher magnification of boxed area in (a). The biocytin-labelled terminal establishes what appears to be a synaptic contact (arrowheads) with the cell body of the reelin-immunoreactive CR cell. Scale bar: 0.2 μm.

unable to recognize the outer molecular layer of the dentate gyrus (Del Rio *et al.*, 1997). We concluded from these studies that not the definitive target cells of entorhinal axons, the granule cells, but early generated pioneer neurons, CR cells, are important for target layer recognition. Moreover, this function of CR cells in pathfinding appears to be specific for the entorhinal axons, as layer-specific ingrowth of commissural fibres, supplied by a co-culture of hippocampus, was preserved in the absence of CR cells. This latter experiment also demonstrated that the application of the glutamate agonist domoate did not alter the tissue such that fibre ingrowth in general was prevented.

4. Reelin is not involved in target recognition of entorhinal axons

As mentioned, CR cells synthesize and secrete the glycoprotein reelin. In order to identify the molecules involved in the target recognition of entorhinal axons, we attempted to study the formation of the entorhinal projection under conditions in which reelin function was perturbed. For this, two approaches were chosen. First, the formation of the entorhinal projection in entorhino-hippocampal co-cultures was studied in the presence of a reelin-blocking antibody (Ogawa *et al.*, 1995). Second, the formation of the entorhinal projection was studied by anterograde tracing in reeler mutant mice lacking reelin. Under both conditions, the entorhinal fibres clearly recognized their appropriate target layer, but the projection appeared to be less pronounced when compared to control co-cultures and wild-type animals, respectively (Del Rio *et al.*, 1997; Deller *et al.*, 1999). This effect was found to be due to a reduced branching of entorhinal axons in the mutant and in co-cultures incubated with the blocking reelin antibody (Del Rio *et al.*, 1997). These results showed that reelin is important for the normal branching pattern of entorhinal axons, but nor for target recognition. The increased collateralization of entorhinal fibres in the presence of reelin was found to be accompanied by an increased number of synapses in the outer molecular layer when compared to mutant tissue (Borell *et al.*, 1999). The question remained as to the nature of the molecules involved in the target recognition of entorhinal axons.

5. Role of CR cells in the pathfinding of entorhinal fibres

The results described above indicate a reelin-independent effect of CR cells in the target layer recognition of entorhinal axons. Moreover, CR cells appear to be involved in the pathfinding of entorhinal axons, because CR cell ablation prevented the entorhinal axons from finding their way to the hippocampus. In recent studies aimed at characterizing the functional properties of CR cells by employing infrared video-microscopy combined with the patch-clamp technique, we discovered that CR cells in complex slices of hippocampus and entorhinal cortex gave rise to axon collaterals invading the entorhinal cortex (Ceranik *et al.*, 1999, 2000). We observed this projection of CR cells to the entorhinal cortex also in co-cultures of hippocampus and entorhinal cortex (Ceranik *et al.*, 1999, 2000). Moreover, by tracing the entorhinal projection with the lipophilic dye, DiI, in fixed tissue from early developmental stages of the rat hippocampal formation, we found retrogradely labelled CR cells in the outer molecular layer following tracer application to the entorhinal cortex as early as on embryonic day 17. These findings clearly indicate that CR cells give rise to an early projection to the entorhinal cortex. With the concept that early pioneer projections form a template for later outgrowing fibres, we hypothesize that the entorhinal axons grow along this early formed hippocampo-entorhinal projection of CR cells (Ceranik *et al.*, 1999; 2000). This hypothesis is compatible with our observation that CR cell elimination prevented the layer-specific ingrowth of entorhinal axons.

6. Conclusions

The results summarized in the present report clearly indicate that hippocampal CR cells play an important role in the formation of a major projection to the hippocampus, the projection from the entorhinal cortex. CR cells are important for three steps in the formation of this projection: pathfinding, target recognition, and synapse formation. We hypothesize that an early projection of CR cells to the entorhinal cortex provides a template for the pathfinding of entorhinal axons. Probably by their synaptic contacts with CR cell somata and dendrites, entorhinal axons are kept in their appropriate target layer in the dentate gyrus, the outer molecular layer. CR cells are required as a transient target, because the definitive target cells, the granule cells, are mainly formed postnatally and are not yet available as postsynaptic partners at the time of entorhinal fibre ingrowth during the late embryonic period. The glycoprotein reelin, synthesized and secreted by CR cells, was found to be involved in the branching of entorhinal axons. In the presence of reelin, entorhinal axons give rise to many more collaterals than in reeler mutant mice or in slice cultures incubated with a blocking reelin antibody. An increased branching of entorhinal axons is accompanied by a larger number of synapses in the outer molecular layer in wild-type animals when compared to reeler mutant mice (Borrell *et al.*, 1999). Thus, reelin appears to be crucially involved in synapse formation of entorhinal axons. Further studies will address the molecular interactions between CR cell axons and entorhinal fibres during the formation of the entorhino–hippocampal projection.

Acknowledgements

Part of these studies was carried out in close collaboration with Dr. E. Soriano's group in Barcelona, including S. Alcantara, V. Borrell, and A. Del Rio. The authors wish to thank R. Hertweck and M. Winter for their help with the illustrations. Our studies contributing to this article were supported by the Deutsche Forschungsgemeinschaft.

References

Borrell, V., Del Rio, J.A., Alcantara, S. *et al.* (1999) Reelin regulates the development and synaptogenesis of the layer-specific entorhino-hippocampal connections. *J. Neurosci.* **19**: 1345–1358.

Ceranik, K., Deng, J., Heimrich, B., Lübke, J., Zhao, S., Förster, E. and Frotscher, M. (1999) Hippocampal Cajal–Retzius cells project to the entorhinal cortex: retrograde tracing and intracellular labelling studies. *Eur. J. Neurosci.* **11**: 4278–4290.

Ceranik, K., Zhao, S. and Frotscher, M. (2000) Development of the entorhino-hippocampal projection: guidance by Cajal–Retzius cell axons. *Ann. N.Y. Acad. Sci.* **911**: 43–54.

D'Arcangelo, G., Miao, G.G., Chen, S.-C., Soares, H.D., Morgan, J.I. and Curran, T. (1995) A protein related to extracellular matrix proteins deleted in the mouse mutant reeler. *Nature* **374**: 719–723.

D'Arcangelo, G., Nakajima, K., Miyata, T., Ogawa, M., Mikoshiba, K. and Curran, T. (1997) Reelin is a secreted glycoprotein recognized by the CR-50 monoclonal antibody. *J. Neurosci.* **17**: 23–31.

Del Rio, J.A., Heimrich, B., Super, H., Borrell, V., Frotscher, M. and Soriano, E. (1996) Differential survival of Cajal–Retzius cells in organotypic cultures of hippocampus and neocortex. *J. Neurosci.* **16**: 6896–6907.

Del Rio, J.A., Heimrich, B., Borrell, V. *et al.* (1997) A role for Cajal–Retzius cells and reelin in the development of hippocampal connections. *Nature* **385**: 70–74.

Deller, T., Drakew, A. and Frotscher, M. (1999) Different primary target cells are important for fiber lamination in the fascia dentata: a lesson from reeler mutant mice. *Exp. Neurol.* **156**: 239–253.

Derer, P. and Derer, M. (1990) Cajal–Retzius cell ontogenesis and death in mouse brain visualized with horseradish peroxidase and electron microscopy. *Neuroscience* **36**: 839–856.

Frotscher, M. (1997) Dual role of Cajal–Retzius cells and reelin in cortical development. *Cell Tissue Res.* **290**: 315–322.

Frotscher, M. (1998) Cajal–Retzius cells, reelin, and the formation of layers. *Curr. Opin. Neurobiol.* **8**: 570–575.

Frotscher, M. and Heimrich, B. (1993) Formation of layer-specific fiber projections to the hippocampus *in vitro*. *Proc. Natl Acad. Sci. USA* **90**: 10400–10403.

Frotscher, M. and Heimrich, B. (1995) Lamina-specific synaptic connections of hippocampal neurons *in vitro*. *J. Neurobiol.* **26**: 350–359.

Hirotsune, S., Takahare, T., Sasaki, N. *et al.* (1995) The reeler gene encodes a protein with an EGF-like motif expressed by pioneer neurons. *Nature Genet.* **10**: 77–83.

Li, D., Field, P.M., Starega, U., Li, Y. and Raisman, G. (1993) Entorhinal axons project to dentate gyrus in organotypic slice co-culture. *Neuroscience* **52**: 799–813.

Meyer, G., Soria, J.M., Martínez-Galán, J.R., Martín-Clemente, B. and Fairén, A. (1998) Different origins and developmental histories of transient neurons in the marginal zone of the fetal and neonatal rat cortex. *J. Comp. Neurol.* **397**: 493–518.

Ogawa, M., Miyata, T., Nakajima, K., Yagyu, K., Seike, M., Ikenaka, K., Yamamoto, H. and Mikoshiba, K. (1995) The reeler gene-associated antigen on Cajal–Retzius neurons is a crucial molecule for laminar organization of cortical neurons. *Neuron* **14**: 899–912.

Rakic, P. and Caviness, V.S. Jr (1995) Cortical development: view from neurological mutants two decades later. *Neuron* **14**: 1101–1104.

Ramón y Cajal, S. (1891) Sur la structure de l'écorce cérébrale de quelques mamifières. *La Cellule* 7: 125–176.

Ramón y Cajal, S. (1911) *Histologie du Système Nerveux de l'Homme et des Vertébrés*, Vol. 2. Maloine, Paris.

Retzius, G. (1893) Die Cajalschen Zellen der Grosshirnrinde beim Menschen und bei Säugetieren. *Biol. Unters.* **5**: 1–9.

Retzius, G. (1894) Weitere Beiträge zur Kenntnis der Cajalschen Zellen der Grosshirnrinde des Menschen. *Biol. Unters.* **6**: 29–34.

Supèr, H. and Soriano, E. (1994) The organization of the embryonic and early postnatal murine hippocampus. II. Development of entorhinal, commissural and septal connections studied with the lipophilic tracer DiI. *J. Comp. Neurol.* **344**: 101–120.

5

Signalling cascade of CNR cadherins (Reelin receptors) in the mammalian brain

T. Yagi

1. Introduction

The central nervous system (CNS) differs in structural organization across species. In particular, the cortex differs considerably in neuronal specialization, laminar differentiation, number and size of cortical areas, the representational patterns of sensory and motor areas, the modular subdivision of areas and the connection of modules and areas. Nevertheless, the mammalian neocortex in each species is a highly ordered structure. The different classes of neurons localize in an organized radial array of six cellular layers ranging from the pial surface to the white matter. Positioning of a particular neuron at a specific cortical layer is an essential step in forming a well-organized neuronal network in the cortex. During spatial positioning, a series of neuronal migrations occur during which postmitotic neurons leave their birthplace in the ventricular zone and move along glial guidance fibres. The molecular events that regulate the interaction of the migrating cells with the environmental cues that dictate their spatial position are not yet completely understood. The recently obtained cadherin-related neuronal receptor (CNR) family could have a significant role in brain formation and function (Yagi and Takeichi, 2000), because CNR family proteins function as multiple receptors for receiving and processing the Reelin signal in building the neocortex. *CNR* family genes are constituted as a gene cluster that is similar to immunoglobulin (*Ig*) and T-cell receptor (*TCR*) gene clusters. Furthermore, differences in the numbers of *CNR* genes containing homologous Reelin-binding domains may be attributable for neocortical variation of neuronal layers and neural networks.

2. CNR family from Fyn tyrosine kinase

A novel diversified cadherin-related neuronal receptor (CNR) family is identified in the mouse CNS using the yeast two-hybrid system with Fyn tyrosine kinase and molecular cloning strategies (Kohmura *et al.*, 1998). Src family containing Fyn forms a

Brain Stem Cells, edited by J. Miyan, M. Thorndyke, P.W. Beesley and C. Bannister.

protein complex with diverse receptors such as membrane-binding IgM and TCRs in the immune system. By analysing Fyn–mice, Fyn has been shown to be crucial for normal brain organization and function. Behavioural abnormalities in Fyn–mice appear as fearful behaviour, spatial learning, suckling behaviour, induction of audiogenic seizure and ethanol sensitivity. Electrophysiological abnormalities occur in induction of long-term potentiation (LTP) in the excitatory synapses and GABAergic synaptic response, and morphological abnormalities are seen in migration of neurons and formation of dendrites (for a review see Yagi, 1999). In neurons, Fyn is enriched in nerve growth cone membranes and the postsynaptic density fraction (Grant *et al.*, 1992).

CNR cadherins contain six extracellular (EC) domains, similar to other cadherins, but have a distinct cytoplasmic domain that interacts with Fyn. The similarity of each CNR at the amino acid level varies from 53% to 80%. Nevertheless, the Arg-Gly-Asp (RGD) motif which is found in most of the ligands of integrins, the characteristic cysteine repeats of the transmembrane and cytoplasmic domains, and PXXP motifs for binding SH3 domain are well conserved in the cytoplasmic domain. Interestingly, 152 amino acids at the C-terminal tail are identical. The cytoplasmic domain bears no similarity to any other proteins in the database. CNR cadherins have different protein features from classic cadherins and protocadherins. CNR1 protein is concentrated in the endoplasmic reticulum and Golgi apparatus but less in the plasma membrane when expressed in L1 and Neuro2A cells; therefore, their homophilic binding activities are not detected. In contrast, synaptic membrane localization of CNR proteins is confirmed by immunohistochemical analyses, even at an electron microscope level (Kohmura *et al.*, 1998). The mechanism for localizing CNR proteins to synaptic membrane is unknown. Both Fyn and CNRs are localized to synaptic junctions and form a co-immunoprecipitated protein complex, suggesting that they may be involved in establishing specific neuronal connections and in signal transduction at the synaptic membrane. *In situ* hybridization studies demonstrate that CNR mRNAs are expressed in various regions of the brain, including neocortex, hippocampus, cerebellum and olfactory bulb. The technique of double-labelling *in situ* reveals that different neurons express distinct sets of *CNR* genes. These properties indicate that CNRs may play a central role in the formation of specific neural connections and circuits in the CNS.

3. Unusual gene structures of CNR family

A striking feature of the eight mouse *CNR* cDNAs is that the nucleotide sequences encoding the N-terminal and transmembrane domains of each CNR are similar but distinct, whereas the nucleotide sequences encoding the C-terminal cytoplasmic domain are identical (Kohmura *et al.*, 1998). It has become apparent that the genomic organization of the CNR family (Pcdhα in human) contains similarities to Ig and TCR gene clusters (Sugino *et al.*, 2000; Wu and Maniatis, 1999). The N-terminal extracellular transmembrane domains, and part of the cytoplasmic tail of each CNR protein, are encoded by a distinct and unusually large first exon. These exons are organized in a tandem array. In contrast, the C-terminal cytoplasmic domain of each CNR protein is identical and is encoded by three small exons located downstream from the cluster of N-terminal exons. Wu and Maniatis termed the tandemly clustered, large exon region of each N-terminal the 'variable region', and the commonly used three exons at the C-terminal the 'constant region'.

In *CNR* genes, three alternative versions of the transcripts have been discovered (Sugino *et al.*, 2000). Each single variable exon has the potential to connect with at least three different C-terminal tails, comparable to class-switching in the Ig heavy chain. Further, conserved nucleotide sequence and gene organization in the *CNR* gene cluster in mice and humans have been discovered. The amino acid sequence from the 'constant region' is identical in mice and humans. Conserved amino acid sequences and orthologous gene order in the 'variable regions' of the *CNR* gene cluster in both mice and humans have also been discovered. However, pseudogenes differ in mice and humans, namely, CNR3 in mouse and Pcdhα 12 in humans are altered into pseudogenes in humans and mice, respectively. The use of different sets of CNR gene clusters may relate to different brain organization and function.

4. CNR family as Reelin receptors

EC1 domain among the eight CNR proteins (Kohmura *et al.*, 1998) are well-conserved (the EC1 domain of classical cadherins is required for specific homophilic interaction (Takeichi, 1990)). Strikingly, the conserved EC1 domain of CNRs is able to bind to Reelin protein (Senzaki *et al.*, 1999). Reelin, a large extracellular matrix protein, is generated from Cajal–Retzius neurons and its function is regulation of the cortical layering and positioning of neurons. *CNR*s are expressed in neurons of the cortical layer during cortical development. The ability of CNR to bind reelin is disrupted by CR-50 monoclonal antibody, which is confirmed as an inhibitory antibody for *in vivo* reelin function, suggesting that CNRs are actual receptors for reelin function. Recent works also reveal that very low density lipoprotein receptor (VLDLR) and apolipoprotein E receptor 2 (ApoER2), which can bind to mammalian Disabled 1 (mDab1) via their cytoplasmic tails, bind to Reelin and are concerned with Reelin-induced tyrosine phosphorylation of mDab1 (D'Arcangelo *et al.*, 1999; Hiesberger *et al.*, 1999). mDab1-deficient mice develop a phenotype indistinguishable from Reelin-minus mutant, *reeler* mice (Howell *et al.*, 1997b; Sheldon *et al.*, 1997). mDab1 is tyrosine-phosphorylated by the Src family, including Fyn. The phosphotyrosine level of primary cultured cortical neurons is increased by exogenous Reelin. A monoclonal antibody, anti-RBD (reelin-binding domain), against the CNR EC1 domain can efficiently block the induction of the tyrosine-phosphorylation level of mDab1 by reelin, suggesting that CNRs, Fyn or Src family, mDab1, and VLDLR or ApoER2 may form a receptor complex with exogenous Reelin. In addition, anti-RBD can inhibit the normal migration of cortical neurons in cultured aggregates (Senzaki *et al.*, 1999). Thus CNR cadherins have the unusual feature of heterophilic interaction with Reelin. The mutations *LIS1*, *doublecortin*, *cdk5*, and *p35* cause anomalous cortical lamina. These proteins are associated with cytoskeletal dynamics and may function relative to the signalling cascades during the reorganization of the cytoskeleton downstream of mDab1, Fyn or the CNR family (*Figure 1*).

5. Positive selection model for neuronal layering and positioning

The structural features of the open reading frames of the CNR gene family are highly unusual: while the N-terminal region is variable, the C-terminal region is identical (Kohmura *et al.*, 1998). Recent studies have revealed a striking similarity in the

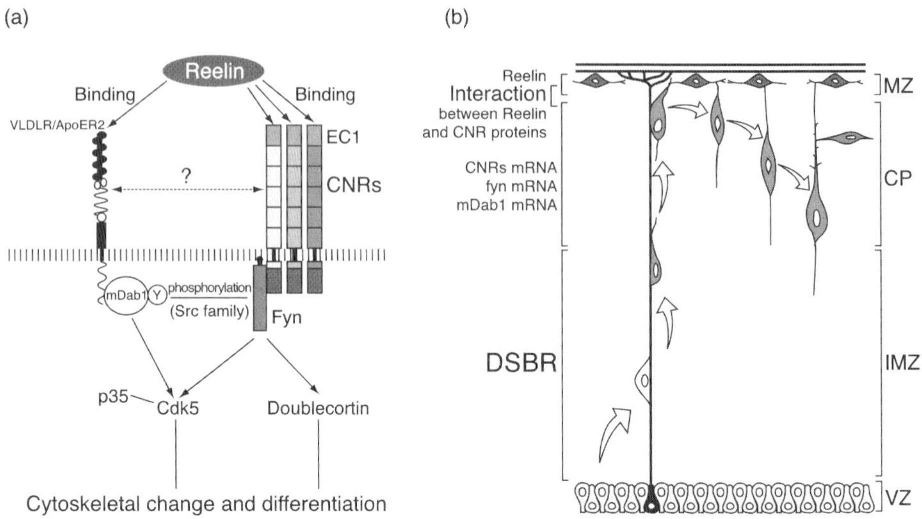

Figure 1. Role of reelin, CNR family, Fyn and mDab1 in cortical plate formation. (a) Hypothetical pathway of reelin signalling. Reelin directly interacts with multiple CNR proteins on the neuronal surface. The CNRs are commonly associated with Fyn-tyrosine kinase. The membrane-bound mDab1 is tyrosine-phosphorylated and induces activation of the downstream components. VLDLR and ApoER2 are necessary for membrane localization of mDab1. After binding to reelin, multiple CNRs induce activation of cytoplasmic components in the cortical neurons. (b) A schematic diagram of neuronal development in the cortex. At postmitotic stage, DNA double–strand break repair reactions (DSBR) in the neurons are necessary for survival. In the cortical surface, migrating neurons contact Cajal–Retzius cells. CNR family proteins expressed in neurons directly bind to reelin expressed by Cajal–Retzius cells. After this interaction, neurons detach from the fibre, develop a terminal dendritic bouquet, grow the apical dendrites, and initiate the synaptogenesis. CNRs, Fyn, and mDab1 are expressed in the cortical layer.

genomic organization of the CNR family genes (*Pcdhα* in humans), the immunoglobulin and T-cell receptor gene clusters. One possible mechanism of gene expression of the CNR family is via the DNA rearrangement model (Wu and Maniatis, 1999). The plausibility of this model in brain development and function is consistent with the recent findings that mice deficient in genes required for double-stranded DNA break repair (XRCC4 and DNA ligase IV) die during embryogenesis due to extensive apoptosis of postmitotic neuronal cells in the cortex (Gao *et al.*, 1998). Interestingly, this cell death correlates with the temporal and spatial pattern of neurogenesis. The embryonic expression patterns of the CNR gene transcripts involve the ‘constant region’ transcripts and correlate relatively well with the time course; at first, transcripts appeared in the postmitotic neurons at a low level and, subsequently, more extensively in the cortical layers. Apropos reelin recognition and the above-mentioned striking genomic organization of the CNR family, the neuronal selection mechanism by CNR gene products would be an intriguing mechanism in the generation of brain formation and function. Nevertheless, as less apoptosis of neurons occurs in *reeler* brains, we believe that neuronal selection by Reelin may determine neuronal differentiation and positioning. As a result, Reelin cues may positively ‘select’ the neurons that possess the potential to express the functional CNR products.

Three models have been proposed to explain the phenotypes of Reelin- and mDab1-deficient mice. One model suggests that Reelin acts as a repellent that induces the cortical preplate to split into the marginal zone and subplate. The failure of the preplate to separate in *reeler* mice causes the cortical plate to develop ectopically underneath the subplate neurons. In another model, a reelin gradient may selectively induce cortical plate neurons to migrate past the subplate neurons. In a third model, Reelin is thought to act as a stop signal that instructs migrating neurons to detach from their glial guidance fibre along which they migrate and proceed with their differentiation program.

The proteins of the CNR family are highly enriched in the marginal zone of the embryonic mouse brain. This indicates that the discovery that the CNR family functions as multiple Reelin receptors is consistent with all of the above models, but the stop signal model is more suitable than the repellent and attractive models. The neuronal expression patterns of the multiple CNR family proteins exhibiting binding to reelin may be regulated during neuronal differentiation. We therefore propose another positive selection model whereby Reelin cues select functional neurons through the expression of CNR family proteins. These cues, via the CNR family, induce neuronal differentiation and positioning by developing a terminal dendritic bouquet, detaching from their glial guidance fibre, growing the apical dendrites, and initiating synaptogenesis (*Figure 1*). The positive selection model is similar to the stop model. This model may be more feasible in light of the previously described features of the CNR family: genetic diversity, binding interactions with reelin, protein structure for cell adhesion, localization in the synaptic junction, and coupling with Fyn-tyrosine kinase. Furthermore, the coupling of CNR proteins to tyrosine kinase signalling and the division of their genomic structure into 'variable' and 'constant' regions is reminiscent of the *Ig* and *TCR* receptor genes. These immunological receptor molecules function in positive and negative selection during the maturation of immune cells; hence, the 'positive selection model' for Reelin may be conceptually broader than the stop model with regard to brain formation or the development of brain function.

6. The role of the CNR family genes in the evolution of brain formation and function

Extant vertebrates are currently comprised of diverse groups, each with diverse and numerous species. The evolution of the *Hox* gene cluster is an intriguing basis for the hypothesis that genomic and morphological complexity are causally linked (Prince *et al.*, 1998). In contrast, the brain differs markedly in structural organization and function across species. Simple changes in the genome may produce profound effects, which give rise to the diversity and complexity of brain structure and function within a species. Such changes can result, for example, in the production of a new type of cell, in lengthening of the proliferation time of an area of the germinal epithelium, in the migration of a set of neurons away from the germline area, or in the radial alignment of the glial cells. The CNR family genes encoding our proposed multiple reelin receptors are arranged in genomic clusters (Sugino *et al.*, 2000; Wu and Maniatis, 1999), which may correspond, respectively, to function in neuronal migration, and differentiation and synaptogenesis during brain development. In evolutionary terms, by probing the Reelin-binding EC1 regions of CNRs, the number of detectable bands is shown to be strikingly different among terrestrial vertebrates. Approximately 20, 10, five, and two bands are found in mammals, birds, reptiles, and amphibians, respectively (Senzaki *et al.*, 1999). In contrast, between one and three bands are commonly detected among these species by probing the cytoplasmic constant region in the CNR genes. The number of stained bands representing the Reelin-binding domains of CNR genes gradually increases from amphibia to mammals in parallel with the transition from brain simplicity to complexity. In contrast, the cytoplasmic constant region remains unchanged among terrestrial vertebrates. In addition, the number of CNR (*Pcdhα*) genes containing the RGD sequence for the interaction with Reelin differ between mouse and human – there are eleven genes in the mouse, but only three in the human. In human CNRs, the RGD motif is changed to HGD, HRD, RGG, RED, or HRD. There is no data available to show whether or not these altered sequences still possess the Reelin-binding activity. Their differing affinities for binding Reelin protein may also influence the layering and positioning of neurons in the human CNS. Thus, the sequence alteration or duplication of the variable region of the CNR gene leads us to question the relatively simple, 'more genes, more complexity' model of evolutionary brain diversification in terms of phenotypic complexity. Therefore, the evolution of the CNR gene cluster provides a particularly intriguing foundation for the hypothesis that genomic and brain complexity are casually linked.

7. Conclusion

The present findings demonstrate a novel role for the CNR gene family in cellular signalling events that extends beyond that of cell adhesion molecules. Several unusual molecular features of the CNR family, including their unusual genomic organization, diverse gene number, conserved EC1 domain, and binding to Fyn-tyrosine kinase, support the framework for the reelin theory of brain structure and function.

Acknowledgements

We thank Dr. K. Senzaki, H. Sugino, S. Hamada, K. Hamaguchi-Hamada for supporting this work. This work was supported by Grants-in-Aid from the Ministry of Education, Science, Sports, and Culture of Japan (TY).

References

D'Arcangelo, G., Homayoui, R., Keshvara, L., Rice, D.S., Sheldon, M. and Curran, T. (1999) Reelin is a ligand for lipoprotein receptors. *Neuron* **24:** 471–479.

Gao, Y., Sun, Y., Frank, K.M. *et al.* (1998) A critical role for DNA end-joining proteins in both lymphogenesis and neurogenesis. *Cell* **95:** 891–902.

Grant, S.G., O'Dell, T.J., Karl, K.A., Stein, P.L., Soriano, P. and Kandel, E.R. (1992) Impaired long-term potentiation, spatial learning, and hippocampal development in fyn mutant mice. *Science* **258:** 1903–1910.

Hiesberger, T., Trommsdorff, M., Howell, B.W., Goffinet, A., Mumby, M.C., Cooper, J.A. and Herz, J. (1999) Direct binding of Reelin to VLDL receptor and ApoE receptor 2 induces tyrosine phosphorylation of disabled-1 and modulates tau phosphorylation. *Neuron* **24:** 481–489.

Howell, B.W., Gertler, F.B. and Cooper, J.A. (1997a) Mouse disabled (mDab1): a Src binding protein implicated in neuronal development. *EMBO J.* **16:** 121–132.

Howell, B.W., Hawkes, R., Soriano, P. and Cooper, J.A. (1997b) Neuronal position in the developing brain is regulated by mouse disabled-1. *Nature* **389:** 733–737.

Kohmura, N., Senzaki, K., Hamada, S., Kai, N., Yasuda, R., Watanabe, M., Ishii, H., Yasuda, M., Mishina, M. and Yagi, T. (1998) Diversity revealed by a novel family of cadherins expressed in neurons at a synaptic complex. *Neuron* **20:** 1137–1151.

Prince, V.E., Joly, L., Ekker, M. and Ho, R.K. (1998) Zebrafish *hox* genes: genomic organization and modified colinear expression patterns in the trunk. *Development* **125:** 407–420.

Senzaki, K., Ogawa, M. and Yagi, T. (1999) Proteins of the CNR family are multiple receptors for Reelin. *Cell* **99:** 635–647.

Sheldon, M., Rice, D.S., D'Arcangelo, G., Yoneshima, H., Nakajima, K., Mikoshiba, K., Howell, B.W., Cooper, J.A., Goldowitz, D. and Curran, T. (1997) Scrambler and yotari disrupt the disabled gene and produce a reeler-like phenotype in mice. *Nature* **389:** 730–733.

Sugino, H., Hamada, S., Yasuda, R., Tuji, A., Matsuda, Y., Fujita, M. and Yagi, T. (2000) Genomic organization of the family of CNR cadherin genes in mice and humans. *Genomics* **63:** 75–87.

Takeichi, M. (1990) Cadherins: a molecular family important in selective cell–cell adhesion. *Annu. Rev. Biochem.* **59:** 237–252.

Wu, Q and Maniatis, T. (1999) A striking organization of a large family of human neural cadherin-like cell adhesion genes. *Cell* 97: 779–790.

Yagi, T. (1999) Molecular mechanisms of Fyn-tyrosine kinase for regulating mammalian behaviors and ethanol sensitivity. *Biochem. Pharmacol.* **57:** 845–850.

Yagi, T. and Takeichi, M. (2000) Cadherin superfamily genes: functions, genomic organization, and neurologic diversity. *Genes Dev.* **14:** 1169–1180.

6

The role of the Reelin pathway in cortical development

G. D'Arcangelo

1. Introduction

In the mammalian central nervous system, the majority of neuronal cells derive from a small population of stem cells that proliferate in the ventricular zone. Initially, progenitor cells divide equally, giving rise to more stem cells (expansion phase). Subsequently, cell division gives rise to daughter cells that exit the cell cycle and become postmitotic (neurogenesis). In the cerebral cortex, committed neuroblasts migrate along radial fibres towards the surface of the brain (migration phase). At the end of their migration, differentiating neurons detach from the radial fibres and acquire their specific phenotype. Birthdating studies demonstrated that cohorts of neurons born at approximately the same time end up at the same level in the radial dimension of the cortex and acquire similar morphologies and chemical phenotypes, giving rise to the characteristic six layers of the mammalian neocortex. The formation of these layers follows an 'inside-out' gradient as younger neurons bypass older cells (Caviness, 1982). Therefore, at the end of neurogenesis, early neuronal populations are found deep in the cortex whereas neurons born towards the end of corticogenesis occupy more superficial layers. The formation of cortical layers depends on the activity of several genes. Some of these genes have been recently discovered because their disruption in naturally occurring or induced mutant mice results in aberrant brain development.

2. The *reeler* mutation

2.1 *The* reeler *phenotype*

A naturally occurring mutant mouse called *reeler* (*rl*) because of its characteristic gait was described several decades ago. Another *reeler* allele named rl^{Orl} also occurred by spontaneous mutation. In the past few years, three more alleles of *reeler* were

Brain Stem Cells, edited by J. Miyan, M. Thorndyke, P.W. Beesley and C. Bannister.

generated by mutagenesis: rl^{tg} (Miao *et al.*, 1994), rl^{Alb1}, and rl^{Alb2} (Flaherty *et al.*, 1992). The phenotype of all known *reeler* alleles is very similar and it has been reviewed in detail (D'Arcangelo and Curran, 1998; Goffinet, 1984).

Reeler is an autosomal recessive mutation. Homozygous mice appear ataxic by the second postnatal week. The locomotor defects are associated with severe hypoplasia of the cerebellum that is likely the major cause of ataxia. At the cellular level, the *reeler* brain exhibits layer disorganization in laminated structures such as the cerebral cortex and the hippocampus. Misplaced neurons are also found outside cortical structures in places such as the facial nucleus, the inferior olivary complex (Goffinet, 1984), the mesencephalic trigeminal nucleus (Terashima, 1996), the nucleus ambiguous (Fujimoto *et al.*, 1998), and the sympathetic preganglionic nucleus in the spinal cord (Yip *et al.*, 2000).

The disruption of cortical layers due to widespread neuronal ectopia is a hallmark of the *reeler* phenotype. Cortical neurons appear able to migrate radially from the ventricular zone and form a cortical plate. However, they do not associate tightly with their cohorts to form cortical layers. Because of this cellular phenotype, it was inferred that the *reeler* gene controls cell positioning at the end of the migratory phase of brain development.

When individual neuronal populations in the developing cerebral cortex were studied in detail, it became clear that the early-born subplate neurons are the cortical cells that are affected the most by the *reeler* mutation. In the mutant, subplate neurons occupy an ectopic superficial layer often referred to as the superplate. Thus, these cells do not segregate away from the other population of early-born neurons, the Cajal–Retzius cells in the marginal zone. The failure of the preplate to split into a superficial marginal zone and a deep subplate layer is the first observable defect in the *reeler* cortex. Later-born cortical plate neurons in *reeler* do not bypass subplate neurons as they do in normal mice, but they accumulate in a deeper location within the cortical plate. The distribution of cortical plate neurons in *reeler* appears to be at random and no distinct layers emerge at the end of the migratory phase. It is not clear whether these cells are unable to bypass subplate neurons or are simply unable to arrange themselves in distinct laminas. It is also apparent that the orientation of apical dendrites is very abnormal and, thus, neurons appear confused with respect to their position within cortex.

In the hippocampus of *reeler* mice, severe cellular ectopia is seen in the pyramidal layer and in the dentate gyrus. The hippocampal pyramidal layer is abnormally split into two or three layers of disoriented cells interrupted by cell-free areas. In the dentate gyrus, granule cells are reduced in number and their dendritic arborizations are grossly misoriented (Stanfield and Cowan, 1979a, b).

The cerebellar hypoplasia is the most readily observable anatomic phenotype in the *reeler* brain. In the adult *reeler* cerebellum the Purkinje cells are slightly reduced in number and are located mostly in deep subcortical clusters. Granule cell number is also dramatically reduced (Goldowitz *et al.*, 1997). The primary defect in cerebellar development may be the absence of the Purkinje cell preplate during embryonic development (Goffinet *et al.*, 1984). It is conceivable that these cells provide trophic factors that act at a short distance and that supports the expansion of the external granular layer in the normal cerebellum. In the *reeler* cerebellum, this factor may be essentially removed from the proliferating granule cells as a result of the Purkinje cell ectopia.

One of the most interesting features of the *reeler* phenotype is that, despite cellular ectopia, the majority of neuronal fibres make appropriate connections on the correct cellular targets. However, axons often travel long distances and follow convoluted paths in order to meet their targets, contributing to the apparent disorganization of cortical structures. One striking example of this axonal projection defect is provided by thalamocortical fibres which, in *reeler* mice, innervate superficial instead of deep cortical layers where their transient targets, the subplate neurons, are located. This observation suggests that connectivity is largely determined by the recognition of specific surface molecules expressed on axonal and target cells. This recognition allows correct connectivity among neuronal classes in the *reeler* brain despite abnormalities in cell positioning. However, cellular ectopia does result in some innervation abnormalities in *reeler*, as reported for synaptic connections on cerebellar Purkinje cells (Mariani *et al.*, 1977). These findings illustrate that correct neuronal positioning is essential for the establishment of appropriate synaptic connectivity and proper brain function.

During cortical development, most cortical neurons migrate radially from the ventricular zone towards the pia along radial fibres. These fibres are processes of neuroepithelial cells that span the entire thickness of the embryonic cortex and terminate by forming end feet in the cell-poor marginal zone. Because of their physical association with migrating neurons, the radial fibres are believed to be important for neuronal migration. In *reeler* mice, the orientation of radial fibres is somewhat oblique rather than perpendicular to the surface, and some end feet appear to terminate prematurely before reaching the pial surface. However, these defects are relatively small compared to the massive and widespread neuronal ectopia. Thus, it is not necessarily sufficient to explain the observed neuronal ectopia. A final, more subtle defect observable in the *reeler* cortex is that the marginal zone is not clearly defined as a cell-poor layer, but many cells invade this layer and are found in close association with the basal lamina. In conclusion, it seems that laminated cortical structures in *reeler* are missing a clue that allows their organization in the radial dimension. The identification of the gene mutated in *reeler* mice provided a better understanding of the molecular events controlling layer formation.

2.2 *The* reeler *gene*

Using a classical genetic strategy, the *reeler* gene was mapped to the proximal region of mouse chromosome 5. But the appearance of a transgenic allele of *reeler*, rl^{tg}, significantly accelerated the discovery of the *reeler* gene coding sequence. This allele originated fortuitously by the integration of a single copy of a transgene into the *reeler* locus (Miao *et al.*, 1994). This allowed the rapid isolation of genomic DNA flanking the transgene that provided a molecular tag for the *reeler* locus. Using this sequence information in a PCR assay, mouse genomic clones were isolated and coding sequences were identified using the exon trapping strategy. The exons were then used as probes to detect the *reeler* gene mRNA and to isolate a full-length cDNA. The gene disrupted in *reeler* was thus cloned and named *reelin (Reln)* (D'Arcangelo *et al.*, 1995). This cDNA contains an open reading frame of 10 383 bp that encodes a large novel protein of 3461 amino acids (*Figure 1*). Shortly after the identification of the cDNA, the genomic structure of the entire mouse *reelin* gene was also completely determined (Royaux *et al.*, 1997b). *reelin* is a large gene composed of 65 exons spread over a region of approximately 450 kb. One

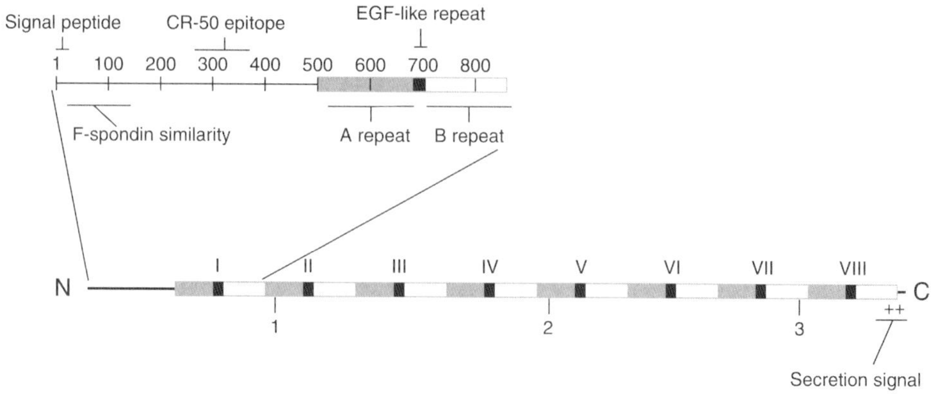

Figure 1. *Structure of the Reelin protein. The Reelin sequence contains an N-terminal region of 500 amino acids followed by eight internal repeats (I-VIII), and a short C terminal region of approximately 30 amino acids. The N-terminus contains a signal peptide, a small region of similarity with F-spondin, and the CR-50 epitope. Each repeat contains 350–390 amino acids and is composed of two related domains (A and B, gray and white boxes, respectively) separated by an EGF-like motif (black box). The C terminus contains a stretch of positively charged amino acids and is required for secretion.*

microexon located near the 3′ end appears to be alternatively spliced, but the significance of this event is unclear (Lambert de Rouvroit *et al.*, 1999a).

The *reelin* gene is highly conserved among vertebrate species but it does not appear to be present in invertebrates such as *Drosophila* or *C. elegans* (DeSilva *et al.*, 1997). The human *reelin* gene is 87.2% identical to the mouse gene at the nucleotide level, suggesting a significant degree of functional conservation between these species. In humans, *reelin* was mapped to chromosome 7q22 (DeSilva *et al.*, 1997). The *reelin* genes from turtle (Bernier *et al.*, 1999), lizard (Goffinet *et al.*, 1999) and chicken (Bernier *et al.*, 2000) have also been recently cloned, allowing speculation on the role of this gene in the evolution of the cortex.

Expression of the *reelin* gene leads to the synthesis of an mRNA of approximately 12 kb which is developmentally regulated in the brain (D'Arcangelo *et al.*, 1995). In adult mice, *reelin* is expressed at lower levels in several tissues but predominantly in the brain. During embryogenesis, *reelin* is expressed mainly in the developing central nervous system, kidney and in the liver (Ikeda and Terashima, 1997). In embryonic cortical structures *reelin* is highly expressed by cells in the marginal zone of the developing neocortex (*Figure 2*) and hippocampus, and by postmitotic neurons in the external granule layer of the cerebellum (D'Arcangelo *et al.*, 1995; Schiffmann *et al.*, 1997). High levels of expression are also detected during early postnatal development in the hippocampus, cerebellum, and retina.

The lack of *reelin* expression in the *rl* and rl^{tg} strains is a consequence of genomic deletions that, in both cases, do not affect the 5′ region of the gene, but presumably result in unstable mRNA (D'Arcangelo *et al.*, 1995). The *rl* breakpoint was further defined to enable the development of PCR-based screening assay that allows the identification of heterozygous and homozygous mice prior to the appearance of the ataxic phenotype (D'Arcangelo *et al.*, 1996). In the rl^{orl} strain, the insertion of an L1 transposable element near the 3′ of the gene causes the skipping of a 220 bp exon (Takahara

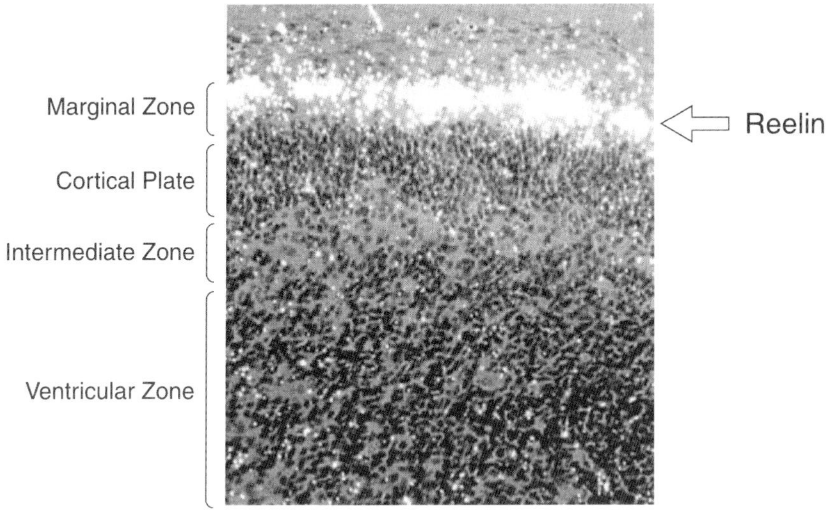

Figure 2. *Reelin expression in the developing cerebral cortex,* reelin *mRNA expression was analyzed by* in situ *hybridization using a* ^{33}*P-labeled antisense probe. The figure represents a darkfield view of a sagittal section of mouse forebrain at embryonic day 13. A strong signal is observed within the marginal zone, but not in other regions of the cortex. The cells that express high levels of* reelin *mRNA are the Cajal–Retzius cells.*

et al., 1996). Even though the levels of *reelin* mRNA are unaffected, this deletion results in an out-of-frame fusion that alters the predicted amino acid sequence of the C-terminus of Reelin (Hirotsune *et al.*, 1995). The chlorambucil-induced *reeler* mutation rl^{Alb2} arose from the insertion of an intracisternal A particle element into the *reeler* locus that, like rl^{orl}, resulted in exon-skipping that causes a frame-shift and is predicted to encode a truncated protein product (Royaux *et al.*, 1997).

2.3 *The Reelin protein*

The *reelin* gene product (Reelin) is a novel protein of 3461 amino acids (approximately 385 kDa) (D'Arcangelo *et al.*, 1995) (*Figure 1*). At the N-terminus, Reelin contains a cleavable signal peptide and a small region of similarity with F-spondin, a protein secreted by floor plate cells in the developing neural tube. At the C-terminus there is a stretch of positively charged amino acids. The main body of the protein comprises a series of eight internal repeats of 350–390 amino acids, each containing two related subdomains that flank a pattern of conserved cysteine residues, known as EGF-like motif. These cysteine-rich regions resemble those found in other extracellular proteins, whereas the flanking subdomains appear to be unique to Reelin. The structural features, together with the presence of potential glycosylation sites, suggested that Reelin is an extracellular protein.

To characterize Reelin further, the full-length protein was expressed *in vitro* and in mammalian cells (D'Arcangelo *et al.*, 1997). The size of the Reelin protein produced *in vitro* is consistent with the predicted molecular weight of 385 kDa. However, the protein appeared larger when analysed from the medium of cultured cells due to the presence of N-linked sugar chains. These studies demonstrated that Reelin is a secreted glycoprotein. Furthermore, it was established that the highly positively charged C-terminal region of Reelin is required for its secretion (D'Arcangelo *et al.*, 1997). Consistently with this finding, it was later demonstrated that Reelin produced by rl^{orl} mice (lacking the C-terminal region) is not secreted (de Bergeyck *et al.*, 1997). Similarly, it can be inferred that the rl^{Alb2} mutation also produces a Reelin protein that cannot be secreted outside the cells. Therefore, the phenotype of all *reeler* alleles known so far is a consequence of the absence of Reelin in the extracellular environment of the developing brain. Recent studies have further shown that Reelin is secreted by a constitutive secretory pathway and is independent of neurotransmitter release (Lacor *et al.*, 2000). After Reelin is synthesized and modified in the Golgi apparatus, it is processed by a protease that has been tentatively identified as a metalloprotease (Lambert de Rouvroit *et al.*, 1999b). This proteolytical event generates two major Reelin fragments of approximately 300 kDa and 180 kDa, in addition to the full-length protein of approximately 400 kDa (D'Arcangelo *et al.*, 1999; Lambert de Rouvroit *et al.*, 1999b). However, the significance of this proteolytic processing is not clear.

The native Reelin protein was studied initially using the monoclonal antibody CR-50. This antibody was raised in *reeler* mice immunized with normal embryonic brain extracts and recognized an antigen present in the brain of normal mice that is absent in *reeler* (Ogawa *et al.*, 1995). The name of the antibody was derived from the fact that it stains Cajal–Retzius cells in the marginal zone of the embryonic cortex. CR-50 is also an interfering antibody, because it was able to recapitulate the *reeler* phenotype *in vitro* in rotating cortical cultures (Ogawa *et al.*, 1995). Further studies demonstrated that CR-50 recognizes Reelin directly and that its epitope lies in the N-terminal region of the Reelin protein (D'Arcangelo *et al.*, 1997). Several studies further demonstrated the ability of the CR-50 antibody to neutralize Reelin activity and mimic the *reeler* phenotype. In explant cultures of embryonic cerebellum, CR-50 disrupts the formation of the Purkinje cell layer (Miyata *et al.*, 1997). In hippocampal slice cultures, CR-50 reduces branching of entorhinohippocampal projections (Del Rio *et al.*, 1997). Finally, when injected into ventricles during embryogenesis, CR-50 disrupts the cellular organization of the pyramidal layer of the hippocampus, resulting in the appearance of cell-poor rifts characteristic of *reeler* mice (Nakajima *et al.*, 1997). More recently, it was demonstrated that the ability of CR-50 to interfere with Reelin function is due to its ability to inhibit the binding of Reelin to its receptors (D'Arcangelo *et al.*, 1999; Senzaki *et al.*, 2000).

Further studies of Reelin were made possible by the development of a panel of monoclonal antibodies (de Bergeyck *et al.*, 1998). One monoclonal antibody, 142, was also useful to study Reelin produced by non-murine species such as humans. Using this antibody, the expression of Reelin in the developing marginal zone of human cortex was described (Meyer and Goffinet, 1998). Furthermore, Reelin was detected for the first time in several human tissues, including adult blood (Smalheisier *et al.*, 2000). Finally, similar immunological studies in primates suggested a role for Reelin in the function of dendritic spines (Rodriguez *et al.*, 2000).

3. The Reelin signalling pathway

3.1 Disabled-1

A phenotype identical to that observed in *reeler* was also described in a spontaneous mouse mutant strain named *scrambler* (*scm*) (Goldowitz *et al.*, 1997; Sweet *et al.*,

1996). The *scm* mutant locus was mapped to chromosome 4, indicating that the affected gene is different from *reelin* and that it may function in the same signalling pathway. The *scm* mutation was identified as a splicing defect in the *disabled-1* (*dab1*) mRNA due to insertion of an intracisternal-A particle in the gene (Sheldon *et al.*, 1997). This defect leads to a significant, but not complete, reduction in the amount of Dab1 protein expressed in the brain of *scm* mice. The *dab1* gene encodes an intracellular protein that is phosphorylated on tyrosine residues during brain development (Howell *et al.*, 1997a, b). Among mammalian genes, *dab1* is mostly similar to *dab2*, the mouse ortholog of the human DOC-2 gene (deleted in ovarian cancer) (Xu *et al.*, 1995). Mammalian *dab1* also displays some similarities to the Drosophila *disabled* gene (*Dab*) (Gertler *et al.*, 1989). In the fly, *Dab* hemizygosity exacerbates the phenotype of Abl–/– mutants, resulting in severe defects in axonal growth. *Dab*-related sequences have also been identified in *C. elegans*, but no function has been reported so far in this organism.

In addition to *scm*, another mouse mutant strain has been described in which the *dab1* mRNA is disrupted by an aberrant splicing mechanism (Sheldon *et al.*, 1997). This strain, named *yotary* (*yot*), arose serendipitously during the generation of an inositol-1,4,5-triphosphate receptor knock-out mouse (Yoneshima *et al.*, 1997). The *yot* mutation, unlike *scm*, results in undetectable levels of Dab1 protein expression (Sheldon *et al.*, 1997). In accordance with studies of spontaneous *dab1* mutants, disruption of the *dab1* gene by homologous recombination also resulted in a phenotype identical to *reeler* (Howell *et al.*, 1997).

Dab1 is an adapter protein that contains a protein interaction/phosphotyrosine-binding motif (PI/PTB) (Margolis, 1996). This motif is also present in another, better know adapter protein named Shc, and it has been implicated in the transduction of growth-inducing signals mediated by the activation of tyrosine kinases. During embryonic brain development, Dab1 is phosphorylated on tyrosine residues and it can interact *in vitro* with several non-receptor tyrosine kinases of the Src family (Howell *et al.*, 1997a). Based on its structure, it was hypothesized that Dab1 may function to mediate Reelin-induced signalling events. In support of this view, Dab1 was found to be expressed in cells adjacent to those producing Reelin, the putative target cells, in all cortical brain structures that are disrupted in *reelin*- or *dab1*-deficient mice (Rice *et al.*, 1998). Dab1 protein levels are also more elevated in *reeler* than in normal brain, suggesting that Reelin may induce its degradation. Conclusive evidence that Dab1 mediates Reelin signalling was provided by recent studies that demonstrated that the addition of Reelin to primary neuronal cultures results in increased tyrosine phosphorylation of Dab1, and that the levels of Dab1 tyrosine phosphorylation are reduced in *reeler* mice (Howell *et al.*, 1999). It is still unclear whether the reduced levels of Dab1 tyrosine phosphorylation in *reeler* mice makes it less susceptible to protein degradation. In conclusion, present data strongly indicate that Dab1 functions downstream of Reelin in a tyrosine kinase signal transduction pathway that controls appropriate cell positioning in the developing brain.

3.2 *Lipoprotein receptors*

Since the discovery of Reln and Dab1, one of the most compelling questions concerning the Reelin signalling pathway has been the identity of cell-surface receptors that mediate the response to Reelin. Recently, this question was answered, thanks in part to the appearance of additional mutant mice that presented with the *reeler* phenotype. By crossing two knock-out mouse strains lacking the very low-density lipoprotein receptor (VLDLR) (Takahashi *et al.*, 1992) and the apolipoprotein E receptor 2 (ApoER2) (Kim *et al.*, 1996), double-mutant mice were generated that lacked both proteins (Trommsdorff *et al.*, 1999). These double-mutant mice are indistinguishable from mice lacking either Reelin or Dab1, whereas individual VLDLR or ApoER2 knock out mice exhibit subtle, but slightly different defects in cell positioning and reduced dendritic arborization (Trommsdorff *et al.*, 1999). The ApoER2 knock-out mice display some lamination defects in the cerebral cortex and hippocampus, although the marginal zone, unlike *reeler*, is clearly defined. VLDLR knock-out mice display obvious defects in the cerebellum. Purkinje cells appear slightly misaligned and present stunted dendrites, although the defect is far less severe than that of *reeler* mice. VLDLR and ApoER2 are members of the low-density lipoprotein receptor (LDLR) family. In mammals, the LDLR family comprises five members, LDLR, LDLR-related protein (LRP), Megalin, VLDLR, and ApoER2. LDLR functions in the systemic clearance of lipoproteins and in the regulation of lipid transport and metabolism (Brown and Goldstein, 1986), whereas the functions of other members of the LDLR family are less well defined. All LDLR family members can also bind apolipoprotein E (apoE)-containing lipoproteins, and some of them also bind and internalize a variety of extracellular ligands, including proteases, protease inhibitors, peptide hormones and vitamin carrier proteins. Despite their nomenclature, it was not at all clear whether VLDL and apoE lipoproteins are true ligands of VLDLR and ApoER2 or whether other ligands for these receptors exist. Particularly in the case of ApoER2, because of the receptor's high and specific expression in the brain, it was thought that brain-specific ligands exist (Kim *et al.*, 1996). Given the high degree of sequence similarity and the phenotype of the double knock-out mutants, it is reasonable to conclude that VLDLR and ApoER2 may have overlapping functions in the brain.

VLDLR and ApoER2 are expressed at relatively high levels throughout the brain, including developing regions affected in the *reeler* mouse (Trommsdorff *et al.*, 1999). These lipoprotein receptors appear to be expressed in the same cell populations that express Dab1 and are targets of Reelin, raising the possibility that VLDLR and ApoER2 may be Reelin receptors. Furthermore, VLDLR and ApoER2, like other members of the LDLR family, can bind Dab1 *in vitro* through its PI/PTB domain (Trommsdorff *et al.*, 1998, 1999). The cytoplasmic domain of the receptors that binds to Dab1 contains an NPxY motif, an amino acid region that is required for receptor internalization (Chen *et al.*, 1990).

Recently, it was demonstrated that Reelin binds directly to VLDLR and ApoER2 *in vivo* (D'Arcangelo *et al.*, 1999) and *in vitro* (Hiesberger *et al.*, 1999). The apparent affinity of Reelin for VLDLR is in the nanomolar range, indicating that Reelin may be a physiological ligand for these lipoprotein receptors in the brain. The interaction of Reelin with VLDLR results in ligand internalization, similarly to other known ligands of LDLR family members (D'Arcangelo *et al.*, 1999). Binding is inhibited by the presence of the interfering antibody CR-50, by excess amounts of apoE, and by RAP, a known non-physiological ligand of lipoprotein receptors (D'Arcangelo *et al.*, 1999; Hiesberger *et al.*, 1999). Calcium chelators, which are known to interfere with the folding of lipoprotein receptor binding domains, also inhibit Reelin binding to VLDLR and ApoER2 (D'Arcangelo *et al.*, 1999; Hiesberger *et al.*, 1999). CR-50, RAP and apoE also interfere with the Reelin-induced stimulation of Dab1 tyrosine phosphorylation in cultured neurons, demonstrating that the binding of Reelin to VLDLR and ApoER2 is essential for the activation of the Reelin signalling pathway (D'Arcangelo *et al.*, 1999; Hiesberger *et al.*, 1999). However, it is unclear how Dab1 phosphorylation is accomplished, since lipoprotein receptors do not possess intrinsic tyrosine kinase activity.

3.3 *Cadherin-like neuronal receptors*

A novel family of genes named cadherin-related neuronal receptors (CNR) was identified on the basis of their interaction with Fyn, a tyrosine kinase of the Src family (Kohmura *et al.*, 1998). These transmembrane proteins were recently shown to also bind Reln *in vitro* with an affinity in the nanomolar range (Senzaki *et al.*, 2000). The binding domain of Reelin was defined by deletion analysis to an N-terminal region that includes the first Reelin subrepeat B, but not the CR-50 epitope. Despite this, CR-50 was shown to interfere with binding of Reelin to CNR proteins. Furthermore, the binding domain of CNR proteins was defined to a region that includes an RGD triple, a motif known to mediate the interaction of other extracellular ligands-receptor partners (Pierschbacher and Ruoslahti, 1984). An antibody directed towards the Reelin binding domain of CNR also interfered with the induction of Dab1 tyrosine phosphorylation and the formation of aggregates in rotating cultures of cortical neurons, two events that are dependent on Reelin function (Senzaki *et al.*, 2000). The discovery that CNRs can function as Reelin receptors is intriguing because these proteins are physically linked to Fyn, a kinase that can phosphorylate Dab1 at least *in vitro* (Howell *et al.*, 1997a). However, it remains to be determined whether Fyn indeed plays a role in Reelin-induced signalling. It is possible that CNR proteins are components of a Reelin receptor complex that includes VLDLR and/or ApoER2, however the phenotype of the double knock-out mice clearly indicates that CNR proteins cannot function alone as Reelin receptors. Finally, it would be interesting to known whether CNR proteins are also necessary for Reelin function *in vivo*. However, given the fact that there are at least eight members of the CNR family, a knock-out approach to answer this question may be arduous.

3.4 *Other potential Reelin-transducing molecules*

The amyloid precursor protein family. The amyloid precursor protein (APP) family includes, in addition to APP, two related genes named APLP1 and APLP2. APP is well known because it is linked to familial Alzheimer's disease (Price and Sisodia, 1998). The function of APLP1 and APLP2 is unknown. All APP family members can bind Dab1 through their cytoplasmic NPxY motif; however, APLP2 binds to a lower extent (Homayouni *et al.*, 1999; Howell *et al.*, 1999b; Trommsdorff *et al.*, 1998). Furthermore, APLP1 and Dab1 co-localize to membrane structures when co-expressed in transfected mammalian cells, and the presence of APLP1 leads to an increase in Dab1 phosphorylation on serine residues (Homayouni *et al.*, 1999). The biological significance of this phosphorylation event remains to be determined. Mice lacking the APP gene do not display ataxia or cortical lamination defects similar to those in *reelin*-, *dab1*- and VLDLR/APOER2-deficient mice (Zheng *et al.*, 1995). Many neuronal functions for APP have been postulated, including differentiation, survival and outgrowth. Thus, it is conceivable that Dab1 participates in these activities, possibly as a modulator of APP family members. Although Reelin does not bind APLP1 (D'Arcangelo *et al.*, 1999), it is also possible that APP family members influence the Reelin pathway by binding Dab1 and changing its phosphorylation state.

Integrins. Integrins mediate the binding of a variety of extracellular ligands by forming receptors composed of many different α and β subunit combinations. In particular, integrin α3 forms a complex with the β1 subunit in migrating neurons. Mice lacking the α3, but not the α5 subunit exhibit neuronal migration defects in the cerebral cortex that are reminiscent, but not identical to that of *reeler* mice (Anton *et al.*, 1999). Interestingly, the integrin β1 subunit also contains NPxY motifs capable of binding Dab1; however, they do not mediate internalization, but rather contribute to the targeting of integrin receptors to focal adhesion sites (Vignaud *et al.*, 1997). There, integrins activate non-receptor tyrosine kinases, which in turn cause cytoskeletal reorganization (Clark and Brugge, 1995). It is conceivable, although not proven, that integrin receptors may be components of a receptor complex that binds Reelin and that they may mediate Dab1 tyrosine phosphorylation through interaction with focal adhesion kinases.

Cdk5. Disruption of either the gene encoding cyclin-dependent kinase 5 (cdk5), a serine–threonine protein kinase, or its activating subunit p35, by homologous recombination results in abnormal positioning of neurons in the cerebral cortex (Chae *et al.*, 1997; Ohshima *et al.*, 1996). The cerebellum and hippocampus of cdk5-null mice are also disrupted similarly to *reeler* mice. These findings raised the possibility that cdk5 and p35 may participate in the Reelin signalling. However, many differences between *reeler* mice and those lacking cdk5 and 35 exist (Gilmore *et al.*, 1998; Kwon and Tsai, 1988). First of all, the lack of cdk5 activity is more severe than that of Reelin, as cdk5-null mice die shortly after birth. p35-null mice also differ significantly from *reeler* mice in that they do not display obvious hippocampal or cerebellar abnormalities and they are not ataxic. Because of its limited expression, p35 could be important for neuronal positioning in the cerebral cortex, but not in the hippocampus or the cerebellum, where other cdk5-activating subunits may replace its function. Both cdk5 and p35 are expressed in cortical neurons in the developing cerebral cortex. In cdk5- and p35-null mice, the marginal zone is normal, whereas it is not clearly defined in *reeler* mice. Furthermore, the preplate splits normally as early cortical neurons bypass subplate cells. However, late-born neurons fail to move past the subplate cells. This finding suggests that cdk5 and p35 are not essential for the initial phases of cortical development but they may be important for later stages of layer formation.

Recent studies have suggested a molecular mechanism that may tie the Reelin signal to cdk5/p35 activity. cdk5 activity promotes the dissociation between β-catenin and N-cadherin, resulting in decreased cell adhesion (Kwon *et al.*, 2000). Thus, it is conceivable that excessive neuronal adhesion prevents the normal migration of late cortical neurons in cdk5/p35-null mice. This mechanism would explain why the initial stages of cortical development are unaffected in cdk5/p35-null mice. It can be speculated that Reelin promotes cortical layer formation through an N-cadherin-mediated mechanism that would involve the inhibition of cdk5 activity. However, a direct link between the Reelin pathway and cdk5 activity has not been found.

4. The function of Reelin

The Reelin signalling pathway plays a crucial role in the development of cortical structures by controlling the end of neuronal migration (*Figure 3*). It is possible that Reelin

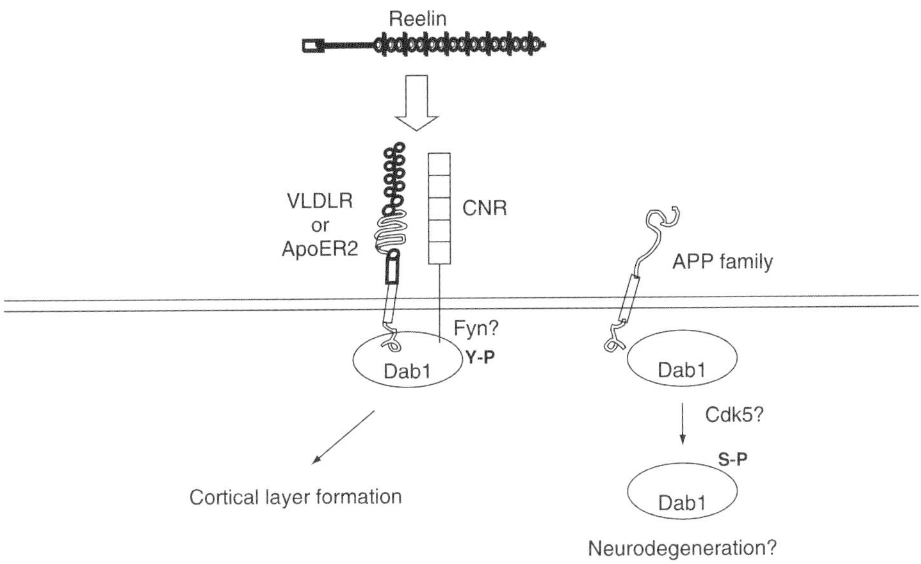

Figure 3. *A model for Reelin Signaling. Extracellular Reelin binds a receptor complex on the surface of migrating neurons. This complex include at least one lipoprotein receptor. VLDLR or ApoER2, or possibly both. The receptor complex may also include CNR proteins. Although the lipoprotein receptors and CNR proteins are capable to bind Reelin individually, the complex may be necessary for Reelin signaling. Dab1 associates with the cytoplasmic domain of lipoprotein receptors. Reelin binding results in increased tyrosine phosphorylation of Dab1 by activation of a tyrosine kinase, possibly Fyn, which may be recruited to the receptor complex by CNR proteins. Through these events, Reelin directs the formation of cortical layers during brain development. Dab1 can also bind APP family members and become phosphorylated by a serine/threonine kinase such as cdk5. This event may play a role in processes related to neuronal degeneration in the adult brain.*

instructs cortical neurons to stop, detach from radial glia fibres, and translocate their cell bodies to a deeper location. This latter event may be prompted by a stimulation of dendritic tree maturation. Reelin could also function as a repulsive signal to subplate neurons, thus favouring the split of the preplate, or as an attractive signal for cortical plate neurons. Finally, Reelin may induce molecular events in cortical plate neurons leading to the expression of adhesion molecules that allow the formation of tight cortical layers. In addition to its role in layer formation, Reelin also influences the maturation of dendritic and axonal processes after the migratory phase is complete.

It is also conceivable that the Reelin pathway plays a role in adult brain function. The interaction between Dab1 and APP family members suggests a role in neurodegeneration. Interestingly, cdk5 immunoreactivity increases in neurons that exhibit early stage neurofibrillary tangles, a hallmark of Alzheimer's disease. cdk5 can phosphorylate Tau, a microtubule-associated protein that accumulates in a hyperphosphorylated form in neurofibrillary tangles (Imahori and Uchida, 1997). cdk5 can also phosphorylate Dab1 efficiently *in vitro* on serine/threonine residues. Finally, *reeler* and VLDLR/ApoER2 double knock-out mice exhibit Tau hyperphosphorylation, suggesting that the binding of Reelin to lipoprotein receptors modulates Tau phosphorylation (Hiesberger *et al.*, 1999). Together, these findings raise the possibility that the Reelin pathway, or some individual components of it, like Dab1, may play a role in pathophysiological processes of neurodegeneration. To address this question more directly, it would be important to determine whether Reelin, Dab1 or VLDLR/ApoER2 mutant mice are more or less susceptible than normal mice to neurodegenerative events.

References

Anton, E.S., Kriedberg, J.A. and Rakic, P. (1999) Distinct functions of alpha3 and alpha5 integrin receptors in neuronal migration and laminar organization of the cerebral cortex. *Neuron* **22**: 277–289.

Bernier, B., Bar, I., Pieau, C., Lambert de Rouvroit, C. and Goffinet, A.M. (1999) Reelin mRNA expression during embryonic brain development in the Turtle *Emys orbicularis*. *J. Comp. Neurol.* **413**: 463–479.

Bernier, B., Bar, I., D'Arcangelo, G., Curran, T. and Goffinet, A.M. (2000) Reelin mRNA expression during embryonic brain development in the chick. *J. Comp. Neurol.* **422**: 488–463.

Brown, M.S. and Goldstein, J.L. (1986) A receptor-mediated pathway for cholesterol homeostasis. *Science* **232**: 34–47.

Caviness, V.S., Jr. (1982) Neocortical histogenesis in normal and reeler mice: a developmental study based upon [^{3}H]thymidine autoradiography. *Brain Res.* **256**: 293–302.

Chae, T., Kwon, Y.T., Bronson, R., Dikkes, P., Li, E. and Tsai, L.H. (1997) Mice lacking p35, a neuronal specific activator of cdk5, display cortical lamination defects, seizures, and adult lethality. *Neuron* **18**: 29–42.

Chen, W.J., Goldstein, J.L. and Brown, M.S. (1990) NPXY, a sequence often found in cytoplasmic tails, is required for coated pit-mediated internalization of the low density lipoprotein receptor. *J. Biol. Chem.* **265**: 3116–3123.

Clark, E.A. and Brugge, J.S. (1995) Integrins and signal transduction pathways: the road taken. *Science* **268**: 233–239.

D'Arcangelo, G. and Curran, T. (1998) *Reeler*: new tales on an old mutant mouse. *BioEssays* **20**: 235–244.

D'Arcangelo, G., Miao, G.G., Chen, S.C., Soares, H., Morgan, J.I. and Curran, T. (1995) A protein related to extracellular matrix proteins deleted in the mouse mutant reeler [see comments]. *Nature* **374**: 719–723.

D'Arcangelo, G., Miao, G.G. and Curran, T. (1996) Detection of the *reelin* breakpoint in reeler mice. *Mol. Brain Res.* **39**: 234–236.

D'Arcangelo, G., Nakajima, K., Miyata, T., Ogawa, M., Mikoshiba, K. and Curran, T. (1997) Reelin is a secreted glycoprotein recognized by the CR-50 monoclonal antibody. *J. Neurosci.* **17**: 23–31.

D'Arcangelo, G., Homayouni, R., Keshvara, L., Rice, D.S., Sheldon, M. and Curran, T. (1999) Reelin is a ligand for lipoprotein receptors. *Neuron* **24**: 471–479.

de Bergeyck, V., Nakajima, K., Lambert de Rouvroit, C., Naerhuyzen, B., Goffinet, A.M., Miyata, T., Ogawa, M. and Mikoshiba, K. (1997) A truncated Reelin protein is produced but not secreted in the "Orleans" reeler mutation ($Reln^{rl\text{-}Orl}$). *Mol. Brain Res.* **50**: 85–90.

de Bergeyck, V., Naerhuyzen, B., Goffinet, A.M. and Lambert de Rouvroit, C. (1998) A panel of monoclonal antibodies against reelin, the extracellular matrix protein defective in reeler mutant mice. *J. Neurosci. Methods* **82**: 17–24.

Del Rio, J. A., Heimrich, B., Borrell, V. *et al.* (1997) A role for Cajal–Retzius cells and reelin in the development of hippocampal connections. *Nature* **385**: 70–74.

DeSilva, U., D'Arcangelo, G., Braden, V.V., Chen, J., Miao, G., Curran, T. and Green, E.D. (1997) The human reelin gene: isolation, sequencing, and mapping on chromosome 7. *Genome Res.* **7**: 157–164.

Flaherty, L., Messer, A., Russell, L.B. and Rinchik, E.M. (1992) Chlorambucil-induced mutations in mice recovered in homozygotes. *Proc. Natl Acad. Sci. USA* **89**: 2859–2863.

Fujimoto, Y., Setsu, T., Ikeda, Y., Miwa, A., Okado, H. and Terashima, T. (1998) Ambiguous nucleus neurons innervating the abdominal esophagus are malpositioned in the reeler mouse. *Brain Res.* **811**: 156–160.
Gertler, F.B., Bennet, R.L., Clark, M.J. and Hoffmann, F.M. (1989) Drosophila *abl* tyrosine kinase in embryonic CNS axons: a role in axogenesis is revealed through dosage-sensitive interactions with *disabled. Cell* **58**: 103–113.
Gilmore, E.C., Ohshima, T., Goffinet, A.M., Kulkarni, A.B. and Herrup, K. (1998) Cyclin-dependent kinase 5-deficient mice demonstrate novel developmental arrest in cerebral cortex. *J. Neurosci.* **18**: 6370–6377.
Goffinet, A.M. (1984) Events governing organization of postmigratory neurons: studies on brain development in normal and reeler mice. *Brain Res.* **319**: 261–296.
Goffinet, A.M., So, K.F., Yamamoto, M., Edwards, M. and Caviness, V., Jr. (1984) Architectonic and hodological organization of the cerebellum in reeler mutant mice. *Brain Res.* **318**: 263–76.
Goffinet, A.M., Bar, I., Trujillo, C., Raynaud, A. and Meyer, G. (1999) Reelin expression during embryonic brain development in lacertian lizards. *J. Comp. Neurol.* **414**: 533–550.
Goldowitz, D., Cushing, R.C., Laywell, E., D'Arcangelo, G., Sheldon, M., Sweet, H.O., Davisson, M., Steindler, D. and Curran, T. (1997) Cerebellar disorganization characteristic of *reeler* in *scrambler* mutant mice despite presence of Reelin. *J. Neurosci.* **17**: 8767–8777.
Hiesberger, T., Trommsdorff, M., Howell, B.W., Goffinet, A.M., Mumby, M.C., Cooper, J. A. and Herz, J. (1999) Direct binding of Reelin to VLDL receptor and ApoE receptor 2 induces tyrosine phosphorylation of Disabled-1 and modulates Tau phosphorylation. *Neuron* **24**: 481–489.
Hirotsune, S., Takahara, T., Sasaki, N. *et al.* (1995) The reeler gene encodes a protein with an EGF-like motif expressed by pioneer neurons. *Nat. Genet.* **10**: 77–83.
Homayouni, R., Rice, D.S., Sheldon, M. and Curran, T. (1999) Dab1 binds to the cytoplasmic domain of amyloid precursor-like protein 1. *J. Neurosci.* **19**: 7507–7515.
Howell, B.W., Gertler, F.B. and Cooper, J.A. (1997a) Mouse disabled (mDab)1): a src binding protein implicated in neuronal development. *EMBO J.* **16**: 121–132.
Howell, B.W., Hawkes, R., Soriano, P. and Cooper, J.A. (1997b) Neuronal position in the developing brain is regulated by mouse disabled-1. *Nature* **389**: 733–736.
Howell, B.W., Herrick, T.M. and Cooper, J.A. (1999a) Reelin-induced tyrosine phosphorylation of Disabled 1 during neuronal positioning. *Genes Dev.* **13**: 643–648.
Howell, B.W., Lanier, L.M., Frank, R., Gertler, F.B. and Cooper, J.A. (1999b) The Disabled-1 PTB domain binds to the internalization signals of transmembrane glycoproteins and to phospholipids. *Mol. Cell Biol.* **19**: 5179–5188.
Ikeda, Y. and Terashima, T. (1997) Expression of *reelin*, the gene responsible for the Reeler mutation, in embryonic development and adulthood in the mouse. *Dev. Dyn.* **210**: 157–172.
Imahori, K. and Uchida, T. (1997) Physiology and pathology of tau protein kinases in relation to Alzheimer's disease. *J. Biochem.* **121**: 179–188.
Kim, D.H., Iijima, H., Goto, K., Sakai, J., Ishii, H., Kim, H. J., Suzuki, H., Kondo, H., Saeki, S. and Yamamoto, T. (1996) Human apolipoprotein E receptor 2. A novel lipoprotein receptor of the low density lipoprotein receptor family predominantly expressed in brain. *J. Biol. Chem.* **271**: 8373–8380.
Kohmura, N., Senzaki, K., Hamada, S., Kai, N., Yasuda, R., Watanabe, M., Ishi, H., Yasuda, M., Mishina, M. and Yagi, T. (1998) Diversity revealed by a novel family of cadherins expressed in neurons at the synaptic complex. *Neuron* **20**: 1137–1151.
Kwon, Y. T. and Tsai, L. H. (1988) A novel disruption of cortical development in p35–/– mice distinct from reeler. *J. Comp. Neurol.* **395**: 510–522.
Kwon, Y. T., Gupta, A., Zhou, Y., Nikolic, M. and Tsai, L.H. (2000) Regulation of N-cadherin-mediated adhesion by the p35/cdk5 kinase. *Curr. Biol.* **10**: 363–372.
Lacor, P.N., Grayson, D.R., Auta, J., Suguya, I., Costa, E. and Guidotti, A. (2000) Reelin secretion from glutamatergic neurons in culture is independent from neurotransmitter regulation. *Proc. Natl Acad. Sci. USA* **97**: 3556–3561.

Lambert de Rouvroit, C., Bernier, B., Royaux, I., de Bergeyck, V. and Goffinet, A.M. (1999a) Evolutionary conserved, alternatively splicing of reelin during brain development. *Exp. Neurol.* **156**: 229–238.

Lambert de Rouvroit, C., de Bergeyck, V., Cortvrindt, C., Bar, I., Eeckhout, Y. and Goffinet, A.M. (1999b) Reelin, the extracellular matrix protein deficient in reeler mutant mice, is processed by a metalloproteinase. *Exp. Neurol.* **156**: 214–217.

Margolis, B. (1996) The PI/PTB domain: a new protein interaction domain involved in growth factor receptor signaling. *J. Lab. Clin. Med.* **128**: 235–241.

Mariani, J., Crepel, F., Mikoshiba, K., Changeux, J.P. and Sotelo, C. (1977) Anatomical, physiological and biochemical studies of the cerebellum from Reeler mutant mouse. *Philos. Trans. R. Soc. Lond. Biol.* **281**: 1–28.

Meyer, G. and Goffinet, A.M. (1998) Prenatal development of Reelin-immunoreactive neurons in the human neocortex. *J. Comp. Neurol.* **397**: 29–40.

Miao, G. G., Smeyne, R.J., D'Arcangelo, G., Copeland, N.G., Jenkins, N.A., Morgan, J.I. and Curran, T. (1994) Isolation of an allele of reeler by insertional mutagenesis. *Proc. Natl Acad. Sci. USA* **91**: 11050–11054.

Miyata, T., Nakajima, K., Mikoshiba, K. and Ogawa, M. (1997) Regulation of Purkinje cell alignment by Reelin as revealed with CR-50 antibody. *J. Neurosci.* **17**: 3599–3609.

Nakajima, K., Mikoshiba, K., Miyata, T., Kudo, C. and Ogawa, M. (1997) Disruption of hippocampal development *in vivo* by CR-50 mAb against Reelin. *Proc. Natl Acad. Sci. USA* **94**: 8196–8201.

Ogawa, M., Miyata, T., Nakajima, K., Yagyu, K., Seike, M., Ikenaka, K., Yamamoto, H. and Mikoshiba, K. (1995) The reeler gene-associated antigen on Cajal–Retzius neurons is a crucial molecule for laminar organization of cortical neurons. *Neuron* **14**: 899–912.

Ohshima, T., Ward, J.M., Huh, C.G., Longenecker, G., Veeranna, G., Pant, H.C., Brady, R.O., Martin, L.J. and Kulkarni, A.B. (1996) Targeted disruption of the cyclin-dependent kinase 5 gene results in abnormal corticogenesis, neuronal pathology and perinatal death. *Proc. Natl Acad. Sci. USA* **93**: 11173–11178.

Pierschbacher, M.D. and Ruoslahti, E. (1984) Variants of the cell recognition site of fibronectin that retain attachment-promoting activity. *Proc. Natl Acad. Sci. USA* **81**: 5985–5988.

Price, D.L. and Sisodia, S.S. (1998) Mutant genes in familial Alzheimer's disease and transgenic models. *Annu. Rev. Neurosci.* **21**: 479–505.

Rice, D.S., Sheldon, M., D'Arcangelo, G., Nakajima, K., Goldowitz, D. and Curran, T. (1998) *Disabled-1* acts downstream of *Reelin* in a signaling pathway that controls laminar organization in the mammalian brain. *Development* **125**: 3719–3729.

Rodriguez, M.A., Pesold, C., Liu, W.S., Khrino, V., Guidotti, A., Pappas, G.D. and Costa, E. (2000) Colocalization of integrin receptors and reelin in dendritic spine postsynaptic densities of adult nonhuman primate cortex. *Proc. Natl Acad. Sci. USA* **97**: 3550–3555.

Royaux, I., Bernier, B., Montgomery, J.C., Flaherty, L. and Goffinet, A.M. (1997a) $Reln^{Alb2}$, an allele of *reeler* isolated from a chlorambucil screen, is due to an IAP insertion with exon skipping. *Genomics* **42**: 479–482.

Royaux, I., Lambert de Rouvroit, C., D'Arcangelo, G., Demirov, D. and Goffinet, A.M. (1997b) Genomic organization of the mouse *reelin* gene. *Genomics* **46**: 240–250.

Schiffmann, S.N., Bernier, B. and Goffinet, A.M. (1997) Reelin mRNA expression during mouse brain development. *Eur. J. Neurosci.* **9**: 1055–1071.

Senzaki, K., Ogawa, M. and Yagi, T. (2000) Proteins of the CNR family are multiple receptors for Reelin. *Cell* **99**: 635–647.

Sheldon, M., Rice, D.S., D'Arcangelo, G., Yoneshima, H., Nakajima, K., Mikoshiba, K., Howell, B.W., Cooper, J.A., Goldowitz, D. and Curran, T. (1997) *Scrambler* and *yotari* disrupt the *disabled* gene and produce a *reeler*-like phenotype in mice. *Nature* **389**: 730–733.

Smalheisier, N.R., Costa, E., Guidotti, A., Impagnatiello, F., Auta, J., Lacor, P., Khrino, V. and Pappas, G.D. (2000) Expression of reelin in adult mammalian blood, liver, pituitary pars intermedia, and adrenal chromaffin cells. *Proc. Natl Acad. Sci. USA* **97**: 1281–1286.

Stanfield, B.B. and Cowan, W.M. (1979a) The development of the hippocampus and dentate gyrus in normal and reeler mice. *J. Comp. Neurol.* **185**: 461–484.

Stanfield, B.B. and Cowan, W.M. (1979b) The morphology of the hippocampus and dentate gyrus in normal and reeler mice. *J. Comp. Neurol.* **185**: 393–422.

Sweet, H.O., Bronson, R.T., Johnson, K.R., Cook, S.A. and Davisson, M.T. (1996) Scrambler, a new neurological mutation of the mouse with abnormalities of neuronal migration. *Mamm. Genome* **7**: 798–802.

Takahara, T., Ohsumi, T., Kuromitsu, J. *et al.* (1996) Dysfunction of the Orleans *reeler* gene arising from exon skipping due to transposition of a full-length copy of an active L1 sequence into the skipped exon. *Hum. Mol. Genet.* **5**: 989–993.

Takahashi, S., Kawarabayasi, Y., Nakai, T., Sakai, J. and Yamamoto, T. (1992) Rabbit very low density lipoprotein receptor: a low density lipoprotein receptor-like protein with distinct ligand specificity. *Proc. Natl Acad. Sci. USA* **89**: 9252–9256.

Terashima, T. (1996) Distribution of mesencephalic trigeminal nucleus neurons in the reeler mutant mouse. *Anat. Rec.* **244**: 563–571.

Trommsdorff, M., Borg, J.P., Margolis, B. and Herz, J. (1998) Interaction of cytosolic adaptor proteins with neuronal apolipoprotein E receptors and the amyloid precursor protein. *J. Biol. Chem.* **273**: 33556–33560.

Trommsdorff, M., Gotthardt, M., Hiesberger, T., Shelton, J., Stockinger, W., Nimpf, J., Hammer, R.E., Richardson, J.A. and Herz, J. (1999) Reeler/Disabled-like disruption of neuronal migration in knockout mice lacking the VLDL receptor and ApoE receptor 2. *Cell* **97**: 689–701.

Vignaud, L., Albiges-Rizo, C., Frachet, P. and Block, M.R. (1997) NPxY motifs control the recruitment of the alpha5beta1 integrin in focal adhesion independently of the association of talin with the beta1 chain. *J. Cell Sci.* **110**: 1421–1430.

Xu, X.X., Yang, W., Jackowski, S. and Rock, C.O. (1995) Cloning of a novel phosphoprotein regulated by colony-stimulating factor 1 shares a domain with the Drosophila disabled gene product. *J. Biol. Chem.* **270**: 14184–14191.

Yip, J.W., Yip, Y.P., Nakajima, K. and Capriotti, C. (2000) Reelin controls position of autonomic neurons in the spinal cord. *Proc. Natl Acad. Sci. USA.* **97**: 8612–8616.

Yoneshima, H., Nagata, E., Matsumoto, M., Yamada, M., Nakajima, K., Miyata, T., Ogawa, M. and Mikoshiba, K. (1997) A novel neurological mutation of mouse, *yotari*, which exhibits *reeler*-like phenotype but expresses *reelin*. *Neurosci. Res.* **29**: 217–223.

Zheng, H., Jiang, M., Trumbauer, M.E. *et al.* (1995) β-Amyloid precursor protein-deficient mice show reactive gliosis and decreased locomotor activity. *Cell* **81**: 525–531.

7

The *Drosophila* nervous system as a model for asymmetric cell division

J.A. Knoblich

1. Introduction

During the development of a multicellular organism, a huge number of different cell types have to be generated from a single precursor cell. In the adult organism, many of these cell types have to be constantly regenerated from undifferentiated cells. This regeneration process is mostly achieved by so-called stem cells which are capable of producing differentiating daughter cells as well as daughter cells that retain stem cell characteristics (Fuchs and Segre, 2000; Weissman, 2000). Both during development and during regeneration any imbalance between different cell types is detrimental to the organism: in the developing nervous system, for example, the right cell types have to be generated at the right time to ensure proper connections. In adult organisms, it is essential to control the size of the stem cell pool to avoid overproliferation and, ultimately, tumorigenesis. Thus, the molecular mechanisms that are used to generate diversity among the progeny of a cell are of tremendous medical importance.

There are several principally different ways in which daughter cells can become different (*Figure 1*) (Horvitz and Herskowitz, 1992). In the most simple case, cell fate determination can be purely stochastic. All daughter cells are initially identical, but then some cells randomly leave the default pathway and assume a different fate (*Figure 1(a)*).

In a more elaborate mechanism (*Figure 1(b)*) this choice is determined by extracellular factors. Even though all daughter cells are initially identical, they experience different concentrations of extracellular factors or come into contact with different molecules on the surface of neighbouring cells and are therefore pushed along different developmental pathways. Growth factors are prominent examples of such extracellular molecules, whereas the Notch-Delta system (Artavanis-Tsakonas *et al.*, 1995) is the best understood cell contact-mediated system that triggers alternative cell fates. Over the past years, however, intrinsically asymmetric cell division has begun to emerge as a third mechanism to generate diversity among the progeny of a cell which

Brain Stem Cells, edited by J. Miyan, M. Thorndyke, P.W. Beesley and C. Bannister.

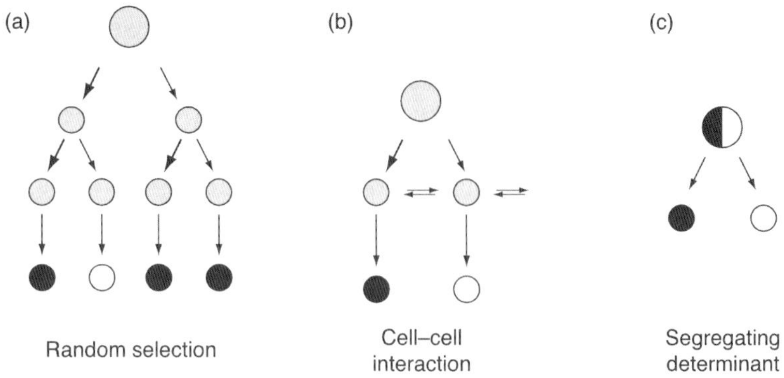

Figure 1. *Mechanisms that can generate different cell fates among the progeny of a single cell. (a) Stochastic selection. (b) External signals. (c) Intrinsically asymmetric cell division.*

relies purely on cell intrinsic signals (*Figure 1(c)*) (Guo and Kemphues 1996; Jan and Jan, 1998, 1999; Knoblich, 1997; Shapiro and Losick, 1997). During such asymmetric cell divisions, cell fate determinants are segregated asymmetrically to one side of a cell before or during mitosis so that different amounts of these determinants are inherited by the two daughter cells. Thus, the two daughter cells are unequal from the time of their birth and these determinants initiate a particular developmental pathway in one of the two daughter cells, but not in the other. Even though such asymmetric cell divisions were postulated many years ago, the first asymmetrically segregating determinant was discovered less than 10 years ago (Rhyu *et al.*, 1994). Until now, asymmetric cell divisions were only well understood in invertebrate model organisms, but the conservation of key factors and their asymmetric segregation in vertebrates, including humans, suggests that similar mechanisms exist in higher organisms.

2. Numb acts as an asymmetrically segregating determinant in the *Drosophila* nervous system

The discovery of the protein Numb in the fruit fly, *Drosophila melanogaster* (Rhyu *et al.*, 1994; Uemura *et al.*, 1989), was a key step in understanding the mechanisms of asymmetric cell division. Numb was identified as a mutation that leads to defects in the development of the *Drosophila* peripheral nervous system (Uemura *et al.*, 1989). Even though subsequent work demonstrated that Numb is involved in the development of a variety of different tissues (Buescher *et al.*, 1998; Carmena *et al.*, 1998; Park *et al.*, 1998; Ruiz Gomez and Bate, 1997; Spana *et al.*, 1995), its function can be most clearly demonstrated during bristle development (*Figure 2*). Bristles are touch receptors that are visible on the back of every adult fly. They are actually external sensory organs and are part of the peripheral nervous system. They are composed of four cells (*Figure 2(a)*): two outer cells, a hair and a socket cell, form the external structures, while two inner cells, the neuron and sheath cell, form internal structures which are not visible from the outside. All four cells arise during pupal development from a single precursor cell, the sensory organ precursor, or SOP cell, in a series of asymmetric cell divisions. The SOP cell divides asymmetrically into an anterior IIa cell and a posterior IIb cell. Later, the IIa cell will form the outer cells, whereas the IIb cell will always form the two

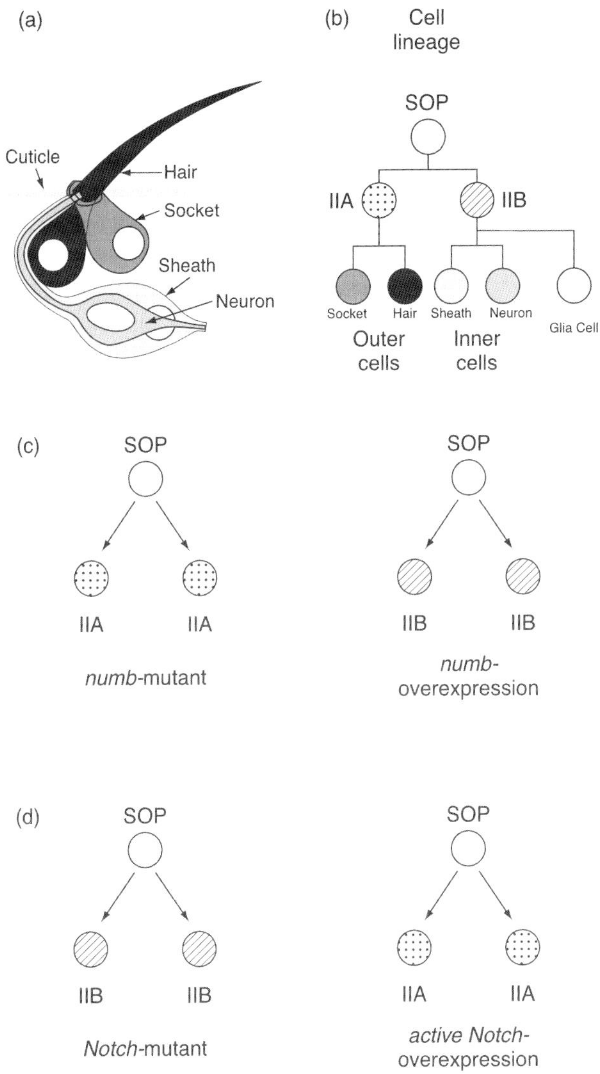

Figure 2. *The function of Numb during bristle development. (a) Bristles are composed of four cells: two outer and two inner cells. These four cells (and an additional glial cell which migrates away) arise during development from a single precursor cell in a series of asymmetric cell divisions. (b) In numb mutants, SOP cells produce two IIa cells, whereas overexpression of numb leads to the opposite cell fate transformation. (c) In the absence of Notch, SOP cells produce two IIb cells, whereas overexpression of activated Notch leads to the opposite cell fate transformation.*

inner cells. In addition, the IIa cell gives rise to a glial cell which migrates away and does not become part of the external sensory (ES) organ.

The gene *numb* is required to make this cell division asymmetric. In the absence of Numb (*Figure 2(b)*), the SOP cell produces two identical IIa cells, which then will give rise to four outer cells with no inner cells. Numb is also required during the second cell division, and in *numb* mutants, in fact, four socket cells are produced, a cell fate transformation that leads to a characteristic visible change in ES organ morphology. When the *numb* gene is overexpressed in SOP cells and their progeny, the opposite cell fate transformation is observed: the SOP cell forms two IIb cells, which then give rise to four inner cells and no outer cells. Since these inner cells are not visible from the outside, this transformation becomes obvious as a bristle-loss phenotype. Thus, Numb acts in making the two daughter cells of the SOP cell different and correct *numb* gene dosage is essential for generating asymmetry in the bristle lineage.

The structure of the Numb protein does not reveal any informative sequence homologies except a phosphotyrosine (PTB)-binding domain at the N-terminus of the protein (Bork and Margolis, 1995; Knoblich *et al.*, 1997). The generation of a Numb antibody, however, revealed a very intriguing subcellular localization that explains Numb function (Rhyu *et al.*, 1994): Numb is associated with the cell membrane and is uniformly distributed around the cortex in interphase cells. In mitotic SOP cells, however, the Numb protein relocalizes during late prophase and concentrates in the area of the cell cortex that overlies one of the two spindle poles. This asymmetric localization is maintained throughout mitosis, and upon cytokinesis the protein segregates preferentially into the IIb cell. Taken together, these facts suggest that Numb induces IIb cell fate in one of the two daughter cells of the SOP and that different levels of Numb generate the difference between the IIa and the IIb cell. Asymmetric localization of Numb is not restricted to SOP cells: a similar subcellular localization and function have been shown in neuroblasts (Spana *et al.*, 1995) and ganglion mother cells (Buescher *et al.*, 1998) of the developing central nervous system; in muscle progenitor cells (Carmena *et al.*, 1998; Park *et al.*, 1998; Ruiz Gomez and Bate, 1997); and during development of the malpighian tubules (Wan *et al.*, 2000). Also, Numb function is not restricted to *Drosophila.* Homologues have been found in chicken (Wakamatsu *et al.*, 1999), mouse (Zhong *et al.*, 1996, 1997), rat and humans, and all experiments that have been done up to now suggest that these homologues serve a similar function in these organisms. Thus, Numb seems to be part of a conserved mechanism that is used in a variety of tissues to generate differential fates among the progeny of a precursor cell.

3. Numb acts by repressing Notch signalling in one daughter cell

Apart from Numb, there is a second class of mutants that lead to cell fate transformations in the bristle lineage. Inactivation of the Notch receptor during or shortly after SOP cell division causes cell fate transformations that are opposite to the ones observed in Numb mutants (*Figure 2(c)*) (Hartenstein and Posakony, 1990). In the absence of Notch, the SOP cell produces only inner cells, whereas overexpression of activated Notch results in the generation of only outer cells. *numb*, *Notch* double mutants have the same phenotype as *Notch* single mutants (Guo *et al.*, 1996; Spana and Doe, 1996), and therefore *numb* formally acts upstream of *Notch* as a repressor in one of the two daughter cells. This repressor function was independently demonstrated in other tissues: Strong overexpression of *numb* during *Drosophila* development can cause *Notch* like phenotypes in the developing fly wing (Frise *et al.*, 1996). Furthermore, transfection of *numb* into tissue culture cells that can respond to activation of Notch by translocating a transcription factor into the nucleus inhibits this response, and this inhibition requires domains in Numb that are also necessary for Numb function *in vivo* (Frise *et al.*, 1996). Taken together, these results suggest that Numb inhibits Notch in the IIb cell.

Despite the clear genetic epistasis, it is not clear how Numb inhibits Notch signalling. Notch is a large transmembrane receptor containing a series of epidermal growth factor (EGF) repeats in the extracellular domain and a series of ankyrin repeats in the intracellular domain. Notch is part of a complex signal transduction process that can be regulated at many levels. After translation, the Notch receptor has to undergo a series of proteolytic maturation cleavages before the cleaved receptor appears at the cell surface. The mature receptor can be activated by either of the two ligands, Delta or Serrate, which are present on neighbouring cells. Ligand binding leads to another cleavage within the intracellular domain and releases the C-terminus of the protein. Upon cleavage, the C-terminal fragment translocates into the nucleus and acts by converting the protein Suppressor of Hairless (Su(H)) into a transcriptional activator. Via an unknown feedback mechanism, this reduces the cell's signalling capacity via Delta and Serrate. Since Notch, Delta and Serrate are present on both the IIa and the IIb cell (but also on the surrounding cells), this signalling mechanism leads to competition between the two cells and the cell with higher Notch signalling level will become the IIa cell, while the other cell will assume the IIb cell fate. Numb is thought to bias this competition so that the cell which inherits Numb will always assume the IIb cell fate. Even though the simplicity of this model for *numb* function is appealing, a role for signalling via Delta from the surrounding cells has not been completely ruled out (Broadus and Doe, 1997; Zeng *et al*., 1998).

The molecular mechanism by which Numb inhibits Notch is not completely clear. Numb interacts with the Notch intracellular domain in a two-hybrid assay, but this interaction has never been verified *in vivo*. Numb could act by inhibiting one of the several maturation cleavages of the Notch receptor. Alternatively, Numb could inhibit the cleavage of the intracellular domain after activation or prevent its translocation into the nucleus. The recent observation that endocytosis plays an important role in Notch signalling (Klueg and Muskavitch, 1999; Parks *et al*., 2000), together with the fact that human Numb interacts with a protein involved in endocytosis (Salcini *et al*., 1997), opens the exciting possibility that Numb functions by regulating endocytosis of the Notch receptor.

4. Prospero, Miranda, Staufen and PON co-localize with Numb, but have different functions

Numb is not the only protein that localizes asymmetrically in *Drosophila* neural precursor cells. The homeodomain transcription factor Prospero was identified based on its function in cell fate specification in the central nervous system (Doe *et al*., 1991; Vaessin *et al*., 1991). The central nervous system develops from neuroblasts, which in turn arise by delamination from epithelial cells in the ventral neuroectoderm. Shortly after delamination, they divide asymmetrically into a larger apical daughter cell that retains neuroblast characteristics and a smaller basal ganglion mother cell (GMC). While the apical daughter cell continues to divide in a stem cell-like fashion, the GMC only divides once more and gives rise to two terminally differentiating neurons. In some of these neuroblast divisions, the prospero transcription factor is required in the GMC to turn on GMC-specific genes, like *even-skipped* (*eve*). In the absence of Prospero, GMCs fail to turn on these genes and continue to express neuroblast-specific genes. Like Numb, the Prospero protein localizes asymmetrically in dividing neuroblasts (Hirata *et al*., 1995; Knoblich *et al*., 1995; Spana and Doe, 1995). Prospero transcription is turned on in late interphase neuroblasts when the protein transiently accumulates at the apical cell cortex. Upon entry into mitosis, Prospero translocates to the basal cell cortex and is segregated into the GMC, where it translocates into the nucleus after mitosis. Even though Numb and Prospero localization are independent of each other, the striking co-localization of the two proteins suggests that they are transported by the same protein localization machinery. In contrast to Numb, however, Prospero does not strictly fulfill the criteria for a segregating determinant. In no case does the absence of Prospero lead to a complete cell fate transformation and – in contrast to Numb – overexpression of Prospero or segregation of the protein into both daughter cells has also not been shown to affect cell fate of the daughter cells. Prospero is needed in the GMC right after mitosis and its asymmetric segregation may be a way to increase the efficiency of its function.

The asymmetric localization of Prospero is mediated by binding to the adapter protein Miranda (Ikeshima-Kataoka *et al.*, 1997; Shen *et al.*, 1997). Miranda was identified in a two-hybrid screen for proteins that bind to the domain of Prospero that is required and sufficient for its asymmetric localization. Miranda is a coiled coil protein that is expressed at high levels in neuroblasts. It co-localizes with Prospero throughout the neuroblast cell cycle and is degraded after mitosis around the time when Prospero translocates into the nucleus. Nuclear translocation of Prospero involves release from Miranda, since Prospero stays at the cell cortex in *miranda* alleles in which Miranda fails to be degraded after mitosis (Matsuzaki *et al.*, 1998). In the absence of Miranda, Prospero is uniformly distributed in the cytoplasm and fails to be segregated into the GMC. Thus, Miranda seems to transport Prospero into the GMC during mitosis.

In addition to Prospero, Staufen is a second protein which is transported by Miranda (Matsuzaki *et al.*, 1998; Schuldt *et al.*, 1998; Shen *et al.*, 1998). Staufen is an RNA-binding protein that is required for specification of the anterior–posterior axis during *Drosophila* oogenesis (St Johnston *et al.*, 1991). In addition to oogenesis, the protein is highly expressed in neuroblasts, and antibody staining has revealed that the protein (like Miranda) accumulates at the apical cell cortex during late interphase, translocates to the basal cortex upon entry into mitosis and segregates into the GMC in telophase (Broadus *et al.*, 1998; Matsuzaki *et al.*, 1998; Schuldt *et al.*, 1998; Shen *et al.*, 1998). Staufen localization requires *miranda* and in *miranda* mutants, Staufen is uniformly distributed in the cytoplasm and fails to segregate into one of the two daughter cells. Like in oogenesis, the function of Staufen in neuroblasts is RNA transport. Staufen binds to the 3′-untranslated region of the *prospero* RNA (Schuldt *et al.*, 1998), and fluorescent *in situ* hybridization has revealed that not only Prospero protein, but also Prospero RNA is asymmetrically segregated during asymmetric cell division in neuroblasts (Broadus *et al.*, 1998; Li *et al.*, 1997). Like the Prospero protein, *prospero* RNA accumulates at the apical cell cortex before mitosis, relocates to the basal cortex upon entry into mitosis and is efficiently segregated into the GMC during cytokinesis. Thus, Miranda functions as an adaptor, which transports both Prospero protein and, via Staufen, *prospero* RNA into the GMC.

In contrast to Prospero protein localization, the function of Prospero RNA localization is not clear. RNA localization is not essential since *staufen* mutants are completely viable even though *prospero* RNA is not asymmetrically segregated in these mutants. However, RNA localization becomes essential in the background of hypomorphic *prospero* mutations (Broadus *et al.*, 1998), suggesting that it serves as a backup mechanism that increases the efficiency of asymmetric Prospero localization, but is dispensable in the presence of an efficient protein transport machinery.

Similarly to Prospero, Numb is also transported by a coiled-coil adaptor protein. The protein Partner-of-Numb (Pon) was identified in a two-hybrid screen for proteins that bind to the PTB domain of Numb (Lu *et al.*, 1998). Pon co-localizes with Numb in dividing neuroblasts and SOP cells, and in *pon* mutants, Numb localization is perturbed. In *pon* mutants, Numb localization does not occur in late prophase, but is delayed until late metaphase. Despite this delay, Numb preferentially segregates into the GMC, and cell fate specification in *pon* mutants is largely normal in the nervous system. However, Pon is required for Numb localization during muscle development. Furthermore, ectopic expression of Pon in epithelial cells, which normally divide symmetrically and do not localize Numb asymmetrically during mitosis, results in asymmetric recruitment of Numb to the basal cell cortex during mitosis. Thus, Pon is sufficient to induce asymmetric localization of Numb, but is not strictly required for this process in neuroblasts and SOP cells. In these cells, another mechanism that is not active in muscle progenitor cells seems to function in parallel to Pon.

What is the molecular mechanism that segregates all these proteins into one of the two daughter cells during mitosis? So far, we only can speculate, but a number of cell biological experiments have excluded certain possibilities. Colcemid treatment does not influence Numb or Prospero localization, suggesting that the process is microtubule-independent (Knoblich *et al.*, 1995). In contrast, the actin drug, Latrunculin A, completely abolishes asymmetric localization showing that actin is an essential component of the localization machinery (Broadus and Doe, 1997; Knoblich *et al.*, 1997; Shen *et al.*, 1998). In addition, movies of Pon localization using a Pon-GFP fusion have excluded localized degradation as a localization mechanism (Lu *et al.*, 1999a). The myosin inhibitor 2,3-butanedione monoxime (BDM) interferes with Pon localization suggesting that actin–myosin-dependent protein transport is the basis for determinant segregation (Lu *et al.*, 1999a).

5. Inscuteable directs asymmetric cell division in neuroblasts

During mitosis, Numb, Prospero, Staufen, Miranda and Pon localize asymmetrically into a cortical crescent in both neuroblasts and SOP cells. Despite the different orientations of mitotic spindles in these two cell types relative to the major body axes, these crescents always form over one of the two spindle poles. This correlation suggests a functional connection between spindle orientation and Numb and Prospero localization. However, both processes are independent of each other: Disruption of mitotic spindles using microtubule drugs does not influence asymmetric localization of Numb and Prospero (Knoblich *et al.*, 1995). Conversely, mitotic spindles are correctly oriented in numb and prospero mutants. Thus, spindle orientation and asymmetric protein localization are highly correlated, but independent events.

This tight correlation requires the function of the Inscuteable protein (Kraut *et al.*, 1996; Kraut and Campos-Ortega, 1996). Like Numb and Prospero, Inscuteable is asymmetrically localized in neuroblasts, but it localizes before mitosis and its crescent forms at the cortical area opposite the Numb and Prospero crescents (*Figure 3(a) and (b)*). Inscuteable is not present in epithelial cells, which divide symmetrically and parallel to the surface of the embryo (*Figure 3(b)*). Inscuteable expression starts when neuroblasts are specified and the protein accumulates in a stalk that is left behind in the epithelial cell layer during delamination. In fully delaminated neuroblasts, Inscuteable forms a tight apical cortical crescent, which is maintained throughout mitosis, but disappears in anaphase. After mitosis, Inscuteable cannot be detected, suggesting that the protein is degraded in telophase. In neuroblasts, Inscuteable is required for both correct spindle orientation and for Numb and Miranda localization to the basal cell cortex. In *inscuteable* mutants, mitotic spindles in neuroblasts assume random orientations. Numb, Prospero and Miranda are still asymmetrically localized in the majority of neuroblasts, but their crescents form at random positions around the cell cortex and are no longer correlated with one of the two spindle poles. Conversely, when *inscuteable* is ectopically expressed in epithelial cells, it induces a rotation of the mitotic spindles in these cells (Kraut *et al.*, 1996). Instead of dividing parallel to the epithelial surface, these cells now divide perpendicularly to the epithelial plane. However, Inscuteable is not sufficient to induce asymmetric localization of Numb in these cells, suggesting that at least one other factor required for Numb localization is specific to neuroblasts. Thus, Inscuteable is a key regulator of asymmetric cell division that orients and coordinates spindle orientation and asymmetric protein localization during asymmetric cell division.

Despite its important function in neuroblasts, Inscuteable is not required for all asymmetric cell divisions that involve the segregation of Numb. SOP cells, which do not divide perpendicularly to the epithelial plane but do localize Numb asymmetrically over one of the two spindle poles, do not require Inscuteable for either spindle orientation or asymmetric protein segregation. Another mechanism must direct asymmetric cell division in these cells, suggesting that Inscuteable is specifically required to direct asymmetric cell divisions along the apical-basal axis.

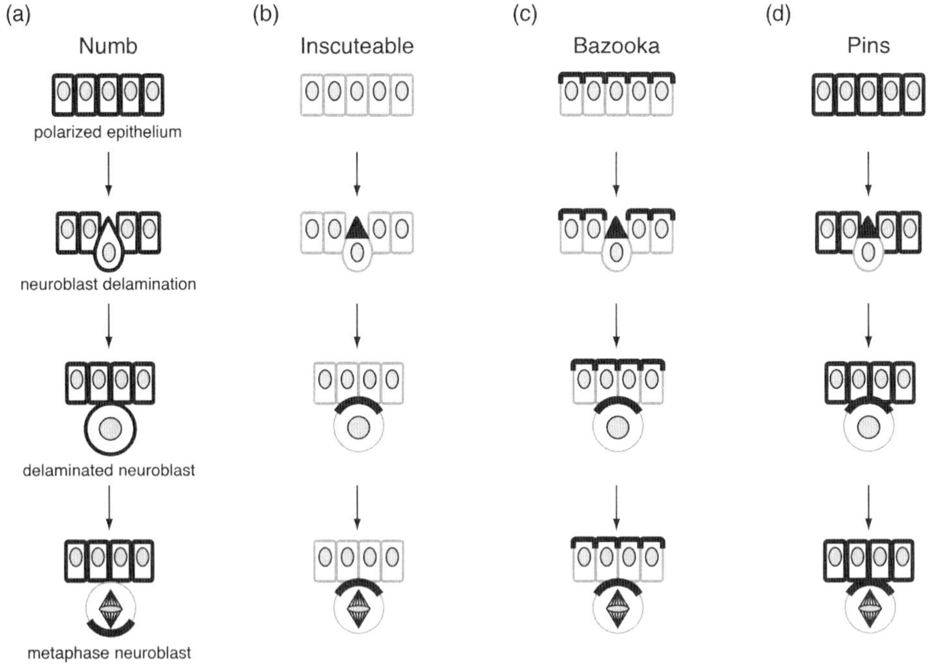

Figure 3. *Subcellular localization of proteins involved in asymmetric cell division in neuroblasts. (a) Numb is homogeneously distributed around the cell cortex in epithelial cells, but localizes asymmetrically in neuroblasts upon entry into mitosis. (b) Inscuteable is not present in epithelial cells. Its expression starts during neuroblast delamination and the protein localizes apically until the end of the first mitosis. (c) Bazooka is localized to the apical cell cortex of both epithelial cells and neuroblasts. (d) Pins is homogeneously distributed around the cell cortex in epithelial cells which do not express inscuteable. In inscuteable-expressing neuroblasts, the protein co-localizes with Inscuteabe at the apical cell cortex.*

6. Bazooka couples asymmetric cell division to epithelial polarity

To ensure that the right cell types are generated at the correct position, asymmetric cell divisions have to be coordinated with the major body axes. Inscuteable orients asymmetric cell division in neuroblasts, but what directs the localization of Inscuteable specifically to the apical cell cortex in these cells? Recent experiments have shown that Inscuteable recognizes certain aspects of epithelial apical–basal polarity that are inherited by neuroblasts during their delamination from the ventral neuroectoderm.

The PDZ domain protein Bazooka, localizes to the apical cell cortex in epithelial cells and is required for maintaining epithelial polarity in these cells (Kuchinke *et al.*, 1998). When neuroblasts delaminate from these epithelial cells, Bazooka does not disappear, but is inherited in the apical stalk that is left behind in the epidermis (*Figure 3(c)*) (Schober *et al.*, 1999; Wodarz *et al.*, 1999). In fully delaminated neuroblasts, Bazooka co-localizes with Inscuteable at the apical neuroblast cell cortex and, like Inscuteable, it becomes undetectable after mitosis. In the absence of Bazooka, epithelial cells have strong defects in apical–basal polarity. Neuroblasts still delaminate and initiate the expression of Inscuteable, but instead of accumulating apically, the protein is distributed in the cytoplasm. As a consequence, mitotic spindles in neuroblasts assume a random orientation and Numb, Prospero and Miranda fail to localize asymmetrically during metaphase. The defects in asymmetric cell division observed in *bazooka* mutants are similar but not identical to those detected in *inscuteable* mutants. While Numb and Miranda still localize asymmetrically in most *inscuteable* mutant neuro-blasts, they fail to do so in the absence of Bazooka, suggesting that Bazooka acts not exclusively through Inscuteable. Thus, Bazooka connects asymmetric cell division in neuroblasts to epithelial apical–basal polarity.

Bazooka encodes the *Drosophila* homologue of Par 3, a protein that directs anterior–posterior polarity of the *C. elegans* zygote suggesting a conserved function in directing cell polarity (Etemad-Moghadam *et al.*, 1995). In *C. elegans*, several other proteins, including the PDZ domain protein, Par-6 (Watts *et al.*, 1996), and the atypical protein kinase C, Pkc-3 (Tabuse *et al.*, 1998), have a function that is very similar to Par-3. Both of these proteins have sequence homologues in *Drosophila* and it will be interesting to see whether a similar group of proteins directs asymmetric cell division in *Drosophila* neuroblasts. Par-3, Par-6 and Pkc-3 are conserved in vertebrates (Izumi *et al.*, 1998; Qiu *et al.*, 2000; Watts *et al.*, 1996) and all experiments done so far suggest a function in cell polarity that is conserved up to humans. Thus, the results obtained in *Drosophila* can be readily transferred to higher organisms and it is intriguing to speculate that a homologous molecular machinery directs asymmetric cell divisions in vertebrates.

7. Frizzled couples asymmetric cell division to planar polarity

While Inscuteable is essential for directing asymmetric cell divisions in *Drosophila* neuroblasts, the protein is not involved in this process in SOP cells. SOP cells divide within the epithelial plane, but a careful analysis of their division planes has revealed a stereotyped orientation of certain SOP cell divisions along the anterior–posterior axis. What is the nature of the polarity that directs these divisions? It has been known for a while that epithelia *in vivo* are polarized not only along the apical–basal axis, but also within the epithelial plane. This phenomenon of 'planar polarity' is particularly well understood in *Drosophila*. One hallmark of tissue polarity mutants is a characteristic misorientation of the bristles on the thorax of a fly. As mentioned above, the cells that form these bristles arise from SOP cells through several asymmetric cell divisions, all of which involve the asymmetric segregation of Numb. A careful phenotypic analysis of planar polarity mutants revealed that they also have defects in the orientation of asymmetric cell divisions (Gho and Schweisguth, 1998). In wild-type flies, SOP cells on the thorax divide with an anterior–posterior orientation and Numb faithfully segregates into the posterior daughter cell, which will become the IIb cell. In flies, mutant for *frizzled*, a seven-transmembrane receptor that is required for planar polarity, the orientations of these divisions are randomized. Similar results were obtained for other tissue polarity mutants, like *dishevelled* and *flamingo* (Gho and Schweisguth, 1998; Lu *et al.*, 1999b).

Thus, while Inscuteable directs asymmetric cell divisions along the apical–basal axis, *frizzled*, *dishevelled* and *flamingo* orient asymmetric cell divisions within the epithelial plane. However, there are characteristic differences: the coordination of the Numb crescent with the mitotic spindle is maintained in *frizzled* mutants, but lost in *inscuteable* mutants. Thus, it is not clear whether these two genes function at the same level of the hierarchy or whether a functional homologue of Inscuteable coordinates Numb localization and spindle orientation in SOP cells. Furthermore, while the SOP cell division requires *frizzled*, subsequent divisions in the SOP lineage always occur with a fixed orientation relative to the first division, suggesting that yet another machinery exists that orients these divisions relative to the SOP division.

8. Pins mediates G-protein signalling and directs asymmetric cell division

While these experiments explain how asymmetric cell divisions, at least in neuroblasts, are coordinated with the body plan, it is completely unclear how apically localized Inscuteable interacts with the spindle orientation machinery and how Inscuteable signals across the cell and directs Numb and Miranda to the basal cell cortex. Several recent experiments suggest that a connection between Inscuteable and heterotrimeric G-proteins may mediate these functions.

The protein, Partner-of-Inscuteable (Pins), was identified in a two-hybrid screen for Inscuteable-binding proteins (Yu *et al.*, 2000), and biochemically as a protein that co-immunoprecipitates with the functional domain of Inscuteable (Schaefer *et al.*, 2000). Pins consists of two domains. The N-terminus is composed of seven TPR repeats and mediates the interaction with Inscuteable. The C-terminus of Pins contains three GoLoco domains (Siderovski *et al.*, 1999), which in other proteins have been shown to bind heterotrimeric G-protein alpha subunits. Indeed, a G-alpha subunit could be identified as a third component of the Pins–Inscuteable complex (Schaefer *et al.*, 2000). Furthermore, Pins was independently identified in a two-hybrid screen for proteins that bind to a heterotrimeric G-protein alpha subunit (Parmentier *et al.*, 2000). The Pins protein is present in both epithelial cells and neuroblasts. While it is located around the cell cortex in epithelial cells (which do not contain Inscuteable), the protein localizes asymmetrically in neuroblasts (*Figure 3(d)*). During delamination, when *inscuteable* expression starts, Pins accumulates in the apical stalk and it co-localizes with Inscuteable throughout neuroblast division. In Inscuteable and Bazooka mutants, Pins fails to localize asymmetrically in neuroblasts. After ectopic expression of Inscuteable in epithelial cells, Inscuteable localizes apically (presumably by binding to Bazooka) and recruits Pins to the apical cell cortex of these cells. Thus, Inscuteable is required and sufficient for Pins apical localization.

The absence of Pins results in an *inscuteable*-like phenotype in neuroblasts: mitotic spindles are misoriented, Numb and Miranda are mislocalized and the correlation between their crescents and spindle orientation is lost. Inscuteable localization is still initiated correctly in these mutants, but the protein fails to be maintained at the apical cell cortex in fully delaminated neuroblasts. Taken together, these results suggest that Inscuteable orients asymmetric cell divisions by apically recruiting Pins. Furthermore, they show that Inscuteable localization is a two-step process, which consists of a Pins-independent initiation phase (in which Inscuteable gets localized by binding to Bazooka) and a Pins-dependent maintenance phase (in which Pins and Inscuteable are co-dependent for asymmetric localization).

The connection between Inscuteable and G-alpha proteins is intriguing. Heterotrimeric G proteins have been implicated in cell polarity in yeast and in *Dictyostelium* (Chant, 1999; Firtel and Chung, 2000). Interestingly, recent experiments have also shown their involvement in orienting mitotic spindles in the early *C. elegans* embryo (Zwaal *et al.*, 1996). A more common function for heterotrimeric G proteins, though, is signal transduction downstream of seven-transmembrane receptors, which in turn are activated by extracellular signals. However, asymmetric localization of both Numb and Inscuteable are cell-autonomous processes which can occur in individually cultured neuroblasts, suggesting that no extracellular ligand is involved (Broadus and Doe, 1997). Interestingly, a vertebrate homologue of Pins was identified based on its capacity to activate heterotrimeric G proteins in the absence of seven-transmembrane receptors (Takesono *et al.*, 1999). It is intriguing to speculate that Inscuteable and Pins function by activating a heterotrimeric G-protein-signalling cascade at the apical, but not basal cell cortex of neuroblasts. The situation could be different in SOP cells: both *flamingo* and *frizzled* encode seven-transmembrane receptors and G proteins could function downstream of these receptors. Thus, heterotrimeric G proteins could be the common readout of *inscuteable*-dependent orientation of asymmetric cell division along the apical–basal axis and *frizzled*-dependent orientation of asymmetric cell divisions along the anterior–posterior axis.

9. Open questions

Despite tremendous progress in our understanding of asymmetric cell division, at least in invertebrate organisms, many questions remain to be solved. What is the molecular mechanism that directs the asymmetric localization of Numb and Miranda? How do Inscuteable and Frizzled interact with this machinery and direct asymmetric cell divisions? In neuroblasts, Inscuteable and Bazooka direct the orientation of the first division along the apical–basal axis. Both proteins disappear after mitosis, yet the second division faithfully repeats the orientation of the first division. How is cell polarity maintained in neuroblasts after the first division?

Despite this progress, our understanding of asymmetric cell divisions in vertebrates, and especially humans, is limited. Most of the players mentioned in this review (with the notable exception of Inscuteable) have homologues in vertebrates. Numb is asymmetrically localized in mouse (Zhong *et al.*, 1996) and chicken (Wakamatsu *et al.*, 1999) and the numb knockout phenotype shows that the protein has an essential function (Zhong *et al.*, 2000). It will be particularly interesting to analyse to what extent the results obtained in lower organisms are applicable to stem cells and to what extent individual stem cells utilize asymmetric cell division to generate diverse cell fates. Recent experiments have demonstrated asymmetric protein segregation in brain stem cells (Johansson *et al.*, 1999), but the functional implications of this asymmetry in protein content remain to be analysed.

References

Artavanis-Tsakonas, S., Matsuno, K. and Fortini, M.E. (1995) Notch signaling. *Science* **268**: 225–232.

Bork, P. and Margolis, B. (1995) A phosphotyrosine interaction domain. *Cell* **80**: 693–694.

Broadus, J. and Doe, C.Q. (1997) Extrinsic cues, intrinsic cues and microfilaments regulate asymmetric protein localization in Drosophila neuroblasts. *Curr. Biol.* **7**: 827–835.

Broadus, J., Fuerstenberg, S. and Doe, C.Q. (1998) Staufen-dependent localization of prospero mRNA contributes to neuroblast daughter-cell fate. *Nature* **391**: 792–795.

Buescher, M., Yeo, S.L., Udolph, G., Zavortink, M., Yang, X., Tear, G. and Chia, W. (1998) Binary sibling neuronal cell fate decisions in the Drosophila embryonic central nervous system are nonstochastic and require inscuteable-mediated asymmetry of ganglion mother cells. *Genes Dev.* **12**: 1858–1870.

Carmena, A., Murugasu-Oei, B., Menon, D., Jimenez, F. and Chia, W. (1998) Inscuteable and numb mediate asymmetric muscle progenitor cell divisions during Drosophila myogenesis. *Genes Dev.* **12**: 304–315.

Chant, J. (1999) Cell polarity in yeast. *Annu. Rev. Cell Dev. Biol.* **15**: 365–391.

Doe, C.Q., Chu-LaGraff, Q., Wright, D.M. and Scott, M.P. (1991) The prospero gene specifies cell fates in the Drosophila central nervous system. *Cell* **65**: 451–464.

Etemad-Moghadam, B., Guo, S. and Kemphues, K.J. (1995) Asymmetrically distributed PAR-3 protein contributes to cell polarity and spindle alignment in early *C. elegans* embryos. *Cell* **83**: 743–752.

Firtel, R.A. and Chung, C.Y. (2000) The molecular genetics of chemotaxis: sensing and responding to chemoattractant gradients. *Bioessays* **22**: 603–615.

Frise, E., Knoblich, J.A., Younger-Shepherd, S., Jan, L.Y. and Jan, Y.N. (1996) The Drosophila Numb protein inhibits signaling of the Notch receptor during cell–cell interaction in sensory organ lineage. *Proc. Natl Acad. Sci. USA* **93**: 11925–11932.

Fuchs, E. and Segre, J.A. (2000) Stem cells: a new lease on life. *Cell* **100**: 143–155.

Gho, M. and Schweisguth, F. (1998) Frizzled signalling controls orientation of asymmetric sense organ precursor cell divisions in Drosophila. *Nature* **393**: 178–181.

Guo, M., Jan, L.Y. and Jan, Y.N. (1996) Control of daughter cell fates during asymmetric division: interaction of Numb and Notch. *Neuron* **17**: 27–41.

Guo, S. and Kemphues, K.J. (1996) Molecular genetics of asymmetric cleavage in the early *Caenorhabditis elegans* embryo. *Curr. Opin. Genet. Dev.* **6**: 408–415.

Hartenstein, V. and Posakony, J.W. (1990) A dual function of the Notch gene in *Drosophila sensillum* development. *Dev. Biol.* **142**: 13–30.

Hirata, J., Nakagoshi, H., Nabeshima, Y. and Matsuzaki, F. (1995) Asymmetric segregation of the homeodomain protein Prospero during Drosophila development. *Nature* **377**: 627–630.

Horvitz, H.R. and Herskowitz, I. (1992) Mechanisms of asymmetric cell division: two Bs or not two Bs, that is the question. *Cell* **68**: 237–255.

Ikeshima-Kataoka, H., Skeath, J.B., Nabeshima, Y., Doe, C.Q. and Matsuzaki, F. (1997) Miranda directs Prospero to a daughter cell during Drosophila asymmetric divisions. *Nature* **390**: 625–629.

Izumi, Y., Hirose, T., Tamai, Y., Hirai, S., Nagashima, Y., Fujimoto, T., Tabuse, Y., Kemphues, K.J. and Ohno, S. (1998) An atypical PKC directly associates and colocalizes at the epithelial tight junction with ASIP, a mammalian homologue of *Caenorhabditis elegans* polarity protein PAR-3. *J. Cell Biol.* **143**: 95–106.

Jan, Y.N. and Jan, L.Y. (1998) Asymmetric cell division. *Nature* **392**: 775–778.

Jan, Y.N. and Jan, L.Y. (1999) Asymmetry across species. *Nat. Cell Biol.* **1**: E42–44.

Johansson, C.B., Momma, S., Clarke, D.L., Risling, M., Lendahl, U. and Frisen, J. (1999) Identification of a neural stem cell in the adult mammalian central nervous system. *Cell* **96**: 25–34.

Klueg, K.M. and Muskavitch, M.A. (1999) Ligand–receptor interactions and trans-endocytosis of Delta, Serrate and Notch: members of the Notch signalling pathway in Drosophila. *J. Cell Sci.* **112**: 3289–3297.

Knoblich, J.A. (1997) Mechanisms of asymmetric cell division during animal development. *Curr. Opin. Cell Biol.* **9**: 833–841.

Knoblich, J.A., Jan, L.Y. and Jan, Y.N. (1995) Asymmetric segregation of Numb and Prospero during cell division. *Nature* **377**: 624–627.

Knoblich, J.A., Jan, L.Y. and Jan, Y.N. (1997) The N terminus of the Drosophila Numb protein directs membrane association and actin-dependent asymmetric localization. *Proc. Natl Acad. Sci. USA* **94**: 13005–13010.

Kraut, R. and Campos-Ortega, J.A. (1996) inscuteable, a neural precursor gene of Drosophila, encodes a candidate for a cytoskeleton adaptor protein. *Dev. Biol.* **174**: 65–81.

Kraut, R., Chia, W., Jan, L.Y., Jan, Y.N. and Knoblich, J.A. (1996) Role of inscuteable in orienting asymmetric cell divisions in Drosophila. *Nature* **383**: 50–55.

Kuchinke, U., Grawe, F. and Knust, E. (1998) Control of spindle orientation in Drosophila by the Par-3-related PDZ-domain protein Bazooka. *Curr. Biol.* **8**: 1357–1365.

Li, P., Yang, X., Wasser, M., Cai, Y. and Chia, W. (1997) Inscuteable and Staufen mediate asymmetric localization and segregation of prospero RNA during Drosophila neuroblast cell divisions. *Cell* **90**: 437–447.

Lu, B., Rothenberg, M., Jan, L.Y. and Jan, Y.N. (1998) Partner of Numb colocalizes with Numb during mitosis and directs Numb asymmetric localization in Drosophila neural and muscle progenitors. *Cell* **95**: 225–235.

Lu, B., Ackerman, L., Jan, L.Y. and Jan, Y.N. (1999a) Modes of protein movement that lead to the asymmetric localization of partner of Numb during Drosophila neuroblast division. *Mol. Cell* **4**: 883–891.

Lu, B., Usui, T., Uemura, T., Jan, L. and Jan, Y.N. (1999b) Flamingo controls the planar polarity of sensory bristles and asymmetric division of sensory organ precursors in Drosophila. *Curr. Biol.* **9**: 1247–1250.

Matsuzaki, F., Ohshiro, T., Ikeshima-Kataoka, H. and Izumi, H. (1998) miranda localizes staufen and prospero asymmetrically in mitotic neuroblasts and epithelial cells in early Drosophila embryogenesis. *Development* **125**: 4089–4098.

Park, M., Yaich, L.E. and Bodmer, R. (1998) Mesodermal cell fate decisions in drosophila are under the control of the lineage genes numb, notch, and sanpodo. *Mech. Dev.* **75**: 117–126.

Parks, A.L., Klueg, K.M., Stout, J.R. and Muskavitch, M.A. (2000) Ligand endocytosis drives receptor dissociation and activation in the Notch pathway. *Development* **127**: 1373–1385.

Parmentier, M.L., Woods, D., Greig, S., Phan, P.G., Radovic, A., Bryant, P. and O'Kane, C.J. (2000) Rapsynoid/Partner of Inscuteable controls asymmetric division of larval neuroblasts in Drosophila. *J. Neurosci. (Online)* **20**: RC84.

Qiu, R.-G., Abo, A. and Martin, G.S. (2000) A human homolog of the *C. elegans* polarity determinant Par-6 links Rac and Cdc42 to PKCζ signaling and cell transformation. *Curr. Biol.* **10**: 697–707.

Rhyu, M.S., Jan, L.Y. and Jan, Y.N. (1994) Asymmetric distribution of numb protein during division of the sensory organ precursor cell confers distinct fates to daughter cells. *Cell* **76**: 477–491.

Ruiz Gomez, M. and Bate, M. (1997) Segregation of myogenic lineages in Drosophila requires numb. *Development* **124**: 4857–4866.

Salcini, A.E., Confalonieri, S., Doria, M., Santolini, E., Tassi, E., Minenkova, O., Cesareni, G., Pelicci, P.G. and Di Fiore, P.P. (1997) Binding specificity and in vivo targets of the EH domain, a novel protein–protein interaction module. *Genes Dev.* **11**: 2239–2249.

Schaefer, M., Shevchenko, A. and Knoblich, J.A. (2000) A protein complex containing inscuteable and the G alpha-binding protein pins orients asymmetric cell divisions in Drosophila. *Curr. Biol.* **10**: 353–362.

Schober, M., Schaefer, M. and Knoblich, J.A. (1999) Bazooka recruits Inscuteable to orient asymmetric cell divisions in Drosophila neuroblasts. *Nature* **402**: 548–551.

Schuldt, A.J., Adams, J.H.J., Davidson, C.M., Micklem, D.R., Haseloff, J., Johnston, D.S. and Brand, A.H. (1998) Miranda mediates asymmetric protein and RNA localization in the developing nervous system. *Genes Dev.* **12**: 1847–1857.

Shapiro, L. and Losick, R. (1997) Protein localization and cell fate in bacteria. *Science* **276**: 712–718.

Shen, C.P., Jan, L.Y. and Jan, Y.N. (1997) Miranda is required for the asymmetric localization of Prospero during mitosis in Drosophila. *Cell* **90**: 449–458.

Shen, C.P., Knoblich, J.A., Chan, Y.M., Jiang, M.M., Jan, L.Y. and Jan, Y.N. (1998) Miranda as a multidomain adapter linking apically localized Inscuteable and basally localized Staufen and Prospero during asymmetric cell division in Drosophila. *Genes Dev.* **12**: 1837–1846.

Siderovski, D.P., Diverse-Pierluissi, M. and De Vries, L. (1999) The GoLoco motif: a Galphai/o binding motif and potential guanine-nucleotide exchange factor. *Trends Biochem. Sci.* **24**: 340–341.

Spana, E.P. and Doe, C.Q. (1995) The prospero transcription factor is asymmetrically localized to the cell cortex during neuroblast mitosis in Drosophila. *Development* **121**: 3187–3195.

Spana, E.P. and Doe, C.Q. (1996) Numb antagonizes Notch signaling to specify sibling neuron cell fates. *Neuron* **17**: 21–26.

Spana, E.P., Kopczynski, C., Goodman, C.S. and Doe, C.Q. (1995) Asymmetric localization of numb autonomously determines sibling neuron identity in the Drosophila CNS. *Development* **121**: 3489–3494.

St Johnston, D., Beuchle, D. and Nusslein-Volhard, C. (1991) Staufen, a gene required to localize maternal RNAs in the Drosophila egg. *Cell* **66**: 51–63.

Tabuse, Y., Izumi, Y., Piano, F., Kemphues, K.J., Miwa, J. and Ohno, S. (1998) Atypical protein kinase C cooperates with PAR-3 to establish embryonic polarity in *Caenorhabditis elegans*. *Development* **125**: 3607–3614.

Takesono, A., Cismowski, M.J., Ribas, C., Bernard, M., Chung, P., Hazard, S., 3rd, Duzic, E. and Lanier, S.M. (1999) Receptor-independent activators of heterotrimeric G-protein signaling pathways. *J. Biol. Chem.* **274**: 33202–33205.

Uemura, T., Shepherd, S., Ackerman, L., Jan, L.Y. and Jan, Y.N. (1989) numb, a gene required in determination of cell fate during sensory organ formation in Drosophila embryos. *Cell* **58**: 349–360.

Vaessin, H., Grell, E., Wolff, E., Bier, E., Jan, L.Y. and Jan, Y.N. (1991) prospero is expressed in neuronal precursors and encodes a nuclear protein that is involved in the control of axonal outgrowth in Drosophila. *Cell* **67**: 941–953.

Wakamatsu, Y., Maynard, T.M., Jones, S.U. and Weston, J.A. (1999) NUMB localizes in the basal cortex of mitotic avian neuroepithelial cells and modulates neuronal differentiation by binding to NOTCH-1. *Neuron* **23**: 71–81.

Wan, S., Cato, A.M. and Skaer, H. (2000) Multiple signalling pathways establish cell fate and cell number in Drosophila malpighian tubules. *Dev. Biol.* **217**: 153–165.

Watts, J.L., Etemad-Moghadam, B., Guo, S., Boyd, L., Draper, B.W., Mello, C.C., Priess, J.R. and Kemphues, K.J. (1996) par-6, a gene involved in the establishment of asymmetry in early *C. elegans* embryos, mediates the asymmetric localization of PAR-3. *Development* **122**: 3133–3140.

Weissman, I.L. (2000) Stem cells: units of development, units of regeneration, and units in evolution. *Cell* **100**: 157–168.

Wodarz, A., Ramrath, A., Kuchinke, U. and Knust, E. (1999) Bazooka provides an apical cue for Inscuteable localization in Drosophila neuroblasts. *Nature* **402**: 544–547.

Yu, F., Morin, X., Cai, Y., Yang, X. and Chia, W. (2000) Analysis of partner of inscuteable, a novel player of Drosophila asymmetric divisions, reveals two distinct steps in inscuteable apical localization. *Cell* **100**: 399–409.

Zeng, C., Younger-Shepherd, S., Jan, L.Y. and Jan, Y.N. (1998) Delta and Serrate are redundant Notch ligands required for asymmetric cell divisions within the Drosophila sensory organ lineage. *Genes Dev.* **12**: 1086–1091.

Zhong, W., Feder, J.N., Jiang, M.M., Jan, L.Y. and Jan, Y.N. (1996) Asymmetric localization of a mammalian numb homolog during mouse cortical neurogenesis. *Neuron* **17**: 43–53.

Zhong, W., Jiang, M.M., Weinmaster, G., Jan, L.Y. and Jan, Y.N. (1997) Differential expression of mammalian Numb, Numblike and Notch1 suggests distinct roles during mouse cortical neurogenesis. *Development* **124**: 1887–1897.

Zhong, W., Jiang, M.M., Schonemann, M.D., Meneses, J.J., Pedersen, R.A., Jan, L.Y. and Jan, Y.N. (2000) Mouse numb is an essential gene involved in cortical neurogenesis. *Proc. Natl Acad. Sci. USA* **97**: 6844–6849.

Zwaal, R.R., Ahringer, J., van Luenen, H.G., Rushforth, A., Anderson, P. and Plasterk, R.H. (1996) G proteins are required for spatial orientation of early cell cleavages in *C. elegans* embryos. *Cell* **86**: 619–629.

8

Developmental abnormalities in early-onset hydrocephalus: clues to signalling

J.A. Miyan, F. Mashayekhi and C.M. Bannister

1. Introduction

Hydrocephalus is a condition defined by an accumulation of cerebrospinal fluid in the fluid spaces of the brain with an associated rise in intracranial pressure. It has long been known that the condition is associated with maldevelopment of the brain, but the precise association between the dilated ventricles, raised pressure and the developmental abnormalities that occur in the condition has not been elucidated. Clinically, not much interest has been taken in the role of the cerebrospinal fluid in hydrocephalus except when it causes life-threatening raised intracranial pressure. The standard method of dealing with this situation is to insert a shunt system to divert fluid from one of the lateral ventricles into the peritoneal cavity. Although this provides an immediate solution to the raised pressure, it now seems likely that the loss of the fluid circulation through the brain may have a major impact on brain development and function. Cerebrospinal fluid is secreted by the choroid plexus and flows through the ventricular system, passing over all regions of germinal activity, and then exits from the fourth ventricle into the basal cisterns from where it passes through the subarachnoid spaces over the surface of the brain and spinal cord. Experimental studies on the hydrocephalic Texas (HTx) rat show that blockage in fluid flow is accompanied by alteration of both neurogenesis and migration of neurons in the developing cerebral cortex before a rise in the intracranial pressure. These data suggest that the cerebrospinal fluid could be an important signalling route for the developing brain and cortex.

2. The fluid system of the central nervous system: the third circulation

In addition to the body's blood and lymphatic systems, a third circulatory system exists in the central nervous system (*Figure 1*) (Nicholson, 1999). Cerebrospinal fluid (CSF) is secreted by the epithelial cells forming the choroid plexus. Choroid plexuses

Brain Stem Cells, edited by J. Miyan, M. Thorndyke, P.W. Beesley and C. Bannister.

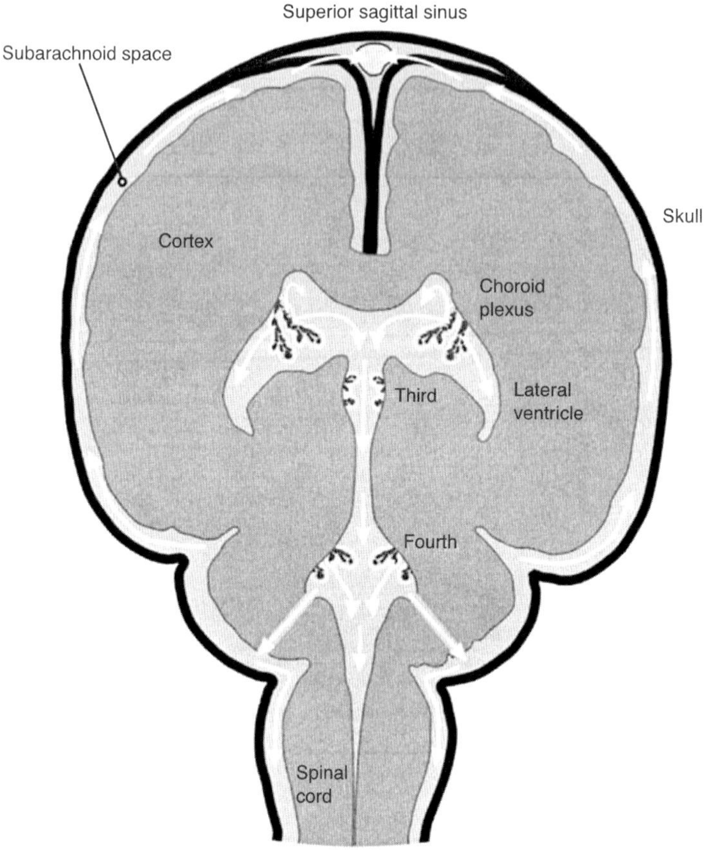

Figure 1. *Diagrammatic representation of the flow of cerebrospinal fluid. CSF is secreted by the choroid plexus located in the lateral, third and fourth ventricles. Fluid flows through the ventricles and exits from the fourth ventricle into the subarachnoid spaces lying over the surface of the brain and spinal cord. Eventually, fluid exits from the CSF pathways into the circulatory system via arachnoid villi into the superior sagittal sinus lying at the top of the cortex between the two cerebral hemispheres.*

are located within the lateral, third and forth ventricles and have a rich capillary network underlying the epithelial sheet.

Fluid flows from the lateral ventricles into the third ventricle. From here it passes through the cerebral aqueduct into the fourth ventricle and then out through the foramina of Lushka and Magendie into the subarachnoid space over the surface of the brain and spinal cord. The fluid is then absorbed into the venous system via the arachnoid villi projecting into the superior sagittal sinus. Other sites of absorbance may be important in addition to the arachnoid villi, but the latter have been identified as the major sites for absorbance of CSF proteins (Mawera and Asala, 1996), suggesting that these proteins must pass over the entire surface of the brain before absorption. In the human CSF is secreted at a rate of 0.35 ml min^{-1} and the fluid spaces of the brain and spinal cord hold 130–140 ml of fluid. This means that approximately four-times the volume of fluid held in the spaces is secreted per day. Fluid secretion and fluid absorption have both passive and active components, with the latter under molecular control. Problems can clearly arise through an imbalance between these two processes that might for example be stimulated by abnormal growth factor concentration (Johanson *et al.*, 1999). The consequences of a blockage in the fluid pathway are also likely to be serious, the seriousness reflecting the severity of restriction in fluid flow and associated rise in pressure as fluid accumulates in proximal compartments. Hydrocephalus is an accumulation of cerebrospinal fluid within the fluid spaces of the brain accompanied by a rise in intracranial pressure. Fluid can accumulate without a persisting rise in pressure and in this circumstance the condition is known as ventriculomegaly since, commonly, the ventricles are enlarged. This condition may precede hydrocephalus but does occur in isolation and is a common occurrence in the human fetus where it is also observed to rectify. Changes in CSF secretion rates and absorbance during development are not known in humans or animals but may be responsible for specific times of onset of fluid accumulation and hydrocephalus in the developing fetus. Studies also show that there are different consequences for sudden versus slow changes in fluid dynamics within the CSF system (Spertell, 1980) suggesting possible strategies for recognition of such conditions.

3. Early-onset hydrocephalus in humans

Human disease is often a spur for research to seek answers to questions that can only be answered in the laboratory. Hydrocephalus is one such case. Much about it has been an enigma for many years. It has long been known that the condition is associated with maldevelopment of the brain but what the precise association is between the dilated ventricles, raised pressure and the developmental abnormalities that occur in the condition has not been elucidated. Clinically, not much thought has been given to the role of the CSF in hydrocephalus and interest is largely confined to its accumulation in its pathways when it causes life threatening raised intracranial pressure. The standard method of dealing with this situation is to insert a shunt system to divert CSF from one of the lateral ventricles into the peritoneal cavity. The brains of premature infants, older children and adults seem to tolerate CSF being drained away from the ventricular system in this manner without any obvious adverse consequences. Because of this, most clinicians accept the commonly held view that the function of the CSF is to act as a cushion for the brain, and even this function is apparently not essential for the post-natal brain's well-being.

Hydrocephalus is not a disease but a condition with many different causes. It can arise at any time of life, occurring in the fetus and throughout postnatal life to late old age. An essential component of hydrocephalus is an increase in the intracranial pressure and this distinguishes it from ventriculomegaly, which is an increase in the size of the ventricular system in the absence of sustained increased pressure. The effect of hydrocephalus on the brain depends on the age of onset, in particular whether the brain is still developing at the time of onset, the cause of the hydrocephalus, the degree of elevation of intracranial pressure and how long the pressure remains raised. Whilst there is no general agreement about the age range for early-onset hydrocephalus, the word 'early' implies that the condition arises during or before the time that the brain is undergoing rapid development. In this paper, early-onset hydrocephalus is defined as hydrocephalus originating in the embryo, the fetus or the prematurely born infant in whom the germinal matrix is known to be still actively producing cells for the developing central nervous system (not just for the olfactory and hippocampus system that continues in the fully developed brain) – that is, at or before the 36th week of human gestation.

Although in any human subject with early onset hydrocephalus the majority of the signs and symptoms exhibited later in life may be due to the root cause of the condition, patients with hydrocephalus of different causes share many common clinical problems; virtually all of them have short-term memory deficits, poor concentration and difficulty in performing sequencing tasks. This suggests that the developing brain, whatever the cause of the hydrocephalus, has been subjected to insults that are specifically related to the presence of the hydrocephalus.

The causes of early-onset hydrocephalus in humans are well documented and are classified as being due to chromosomal abnormalities, genetic defects and acquired disease. Chromosomal abnormalities in general result in such profound and widespread abnormalities throughout the brain and elsewhere in the body that study of their brains is unlikely to contribute to a better understanding of how hydrocephalus affects brain development. Genetic causes have a more restricted effect, but while a number of different genetic defects cause hydrocephalus, in only one, X-linked hydrocephalus, has the abnormal gene been identified (Coucke *et al.*, 1994; Jouet and Kenwrick, 1995; MacFarlane *et al.*, 1997; Van Camp *et al.*, 1993). The cause of hydrocephalus in this condition is due to stenosis of the aqueduct of Sylvius. A number of other causes of hydrocephalus are almost certainly the result of an abnormal gene acting in conjunction with abnormal environmental effects, the best know of these being *spina bifida aperta*, in which a myelomeningocele is associated in nearly every case with a Chiari II malformation of the posterior brain, ventriculomegaly and other abnormalities of the cerebral hemispheres. For reasons that will be discussed later, hydrocephalus generally only develops in these cases after birth. Other causes of hydrocephalus with an inheritance pattern highly suggestive of being due to an abnormal gene are the Dandy-Walker malformation, non-X-linked aqueduct stenosis and communicating hydrocephalus, where the block in the CSF pathways is thought to be in the subarachnoid spaces. The acquired causes of hydrocephalus both pre- and post-natally are viral infections, intraventricular haemorrhages and tumours. As with chromosomal abnormalities, the hydrocephalus in these cases is often accompanied by devastating effect on the whole of the brain by the root cause so that the contribution of hydrocephalus to the abnormal brain maldevelopment is frequently overshadowed by them.

With all the afore-mentioned causes, hydrocephalus is the result of a block somewhere in the CSF pathway. The block may be in the aqueduct, the exit foramina of the fourth ventricle, in the CSF pathways at the level of the foramen magnum, in the subarachnoid spaces over the surface of the brain or in the arachnoid villi. Unfortunately, human brain tissue from both fetuses and neonatal infants with hydrocephalus is seldom available for detailed histological study. Rarely can fresh samples of human brain tissue from these age groups be obtained, so the amount of information in the literature is extremely limited. Most of the information available about the development of the human brain has been collected from studies of fetal brain images and by correlating the abnormalities seen with the long-term outcome. Until recently, imaging has been largely by ultrasound (US) scans, but in the last few years, and increasingly so in the last 2 years with the development of technology allowing fast image capture, it has been possible to visualize the human fetal brain using magnetic resonance (MR) scanning.

Of the above causes of hydrocephalus, the one most likely to yield useful information about the abnormal development of the human brain is genetic abnormality. The brains of fetuses with spina bifida aperta are of particular interest because the onset of hydrocephalus is generally only after birth, and this allows the abnormalities of brain development to be imaged independently of the effect of raised intracranial pressure and the consequent dilatation of the ventricular system. US scans are carried out routinely on all fetuses at 18 to 20 weeks of gestation. In fetuses with spina bifida aperta the skull shape is abnormal and is described as being lemon shaped (*Figure 2*) (Bannister *et al.*, 1998). The back of the lemon is due to the Chiari II malformation. The small cerebellar hemispheres, the downward displacement of the cerebellar tonsils into the upper cervical spinal canal, together with herniation of the lower part of the medulla and the fourth ventricle, results in the posterior fossa of the skull being abnormally small. The front of the lemon is caused by abnormal development of the frontal lobes of the cerebral hemispheres. The lateral ventricles are larger than normal and, particularly over the frontal horns, the cortical mantle is thinner than normal. In these fetuses the downward displacement of the posterior fossa structures through the foramen magnum obstructs the subarachnoid spaces preventing the CSF, which exits from the fourth ventricle into the subarachnoid spaces covering the spinal cord, from re-entering the intracranial cavity. This effectively cuts off the surface of the cerebral cortex from CSF circulation. The central canal of the spinal cord is open in spina bifida aperta and

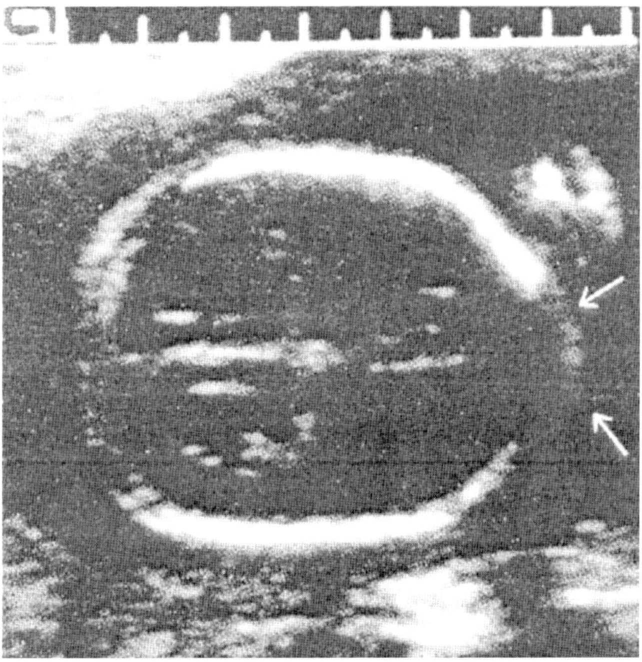

Figure 2. *Ultrasound scan of fetus with spina bifida showing the lemon-shaped cranium.*

connects the cavity of the fourth ventricle to an opening on the surface of the neuroplaque. The central canal of the spinal cord thus acts as an escape route for CSF to leave the ventricular system and prevents hydrocephalus from occurring until the back wound is closed. Closure of the back wound almost inevitably results in hydrocephalus since the CSF no longer has an escape route and the route to the cortical subarachnoid space is thought to remain blocked. As fluid accumulates in the lateral ventricles, the pathophysiological consequences of brain compression and stretching following the increase in the intracranial pressure are observed and can be controlled or even to some extent reversed by fluid drainage with a surgically implanted shunt (Hanigan *et al.*, 1991; Mataro *et al.*, 2000; Oberbauer, 1985; Rekate, 1997).

Two consequences of CSF leaving the ventricular system by the spinal canal–neuroplaque route in spina bifida operta are (i) that intracranial pressure does not rise until the wound is closed, and (ii) that CSF is unable to flow from the ventricles over the cortical surface of the brain. This presents an opportunity to assess the developmental consequence of lack of CSF supply to the cortical subarachnoid space in the absence of raised pressure. MR scans (*Figure 3*) have shown that the subarachnoid

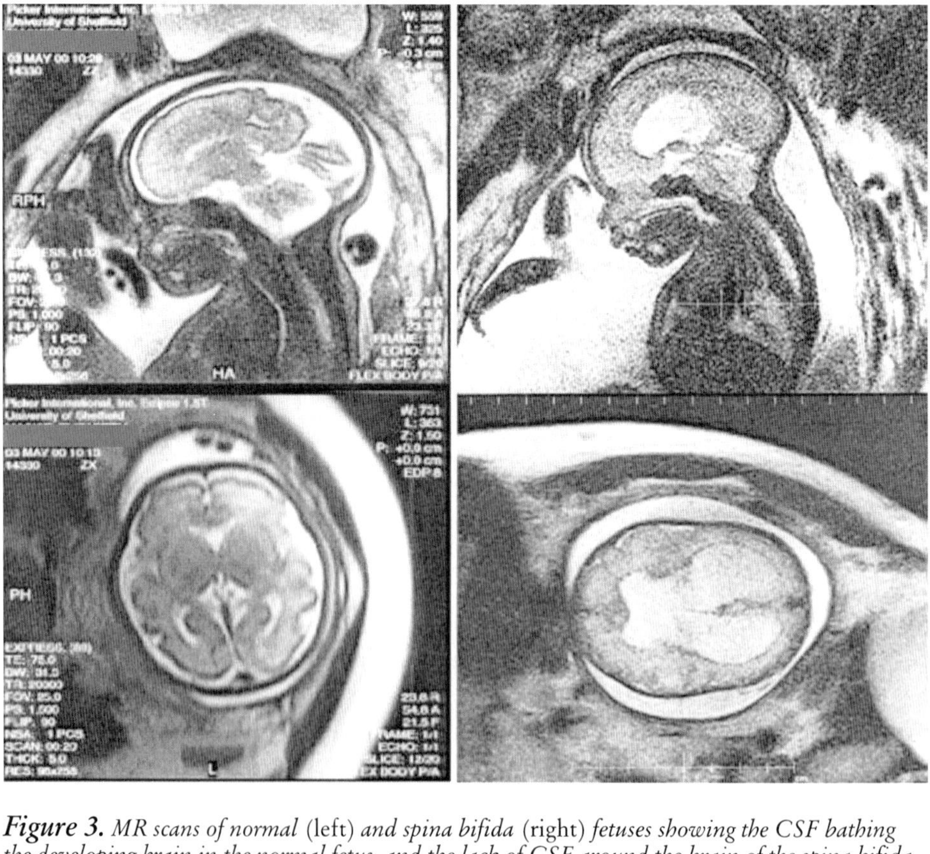

Figure 3. *MR scans of normal* (left) *and spina bifida* (right) *fetuses showing the CSF bathing the developing brain in the normal fetus, and the lack of CSF around the brain of the spina bifida fetus. The enlarged ventricles of the abnormal brain are thought to be due to poor development of the cortex. The child is born with a soft fontanelle indicating no increase in intracranial pressure. Fluid accumulates and fills the space into which the cortex would normally have developed. Upper panels show sagittal MR scans, lower panels show axial (horizontal) scans.*

spaces are prominent in the normal human fetus. This is in marked contrast to the brains of fetuses with spina bifida aperta, where MR scans (*Figure 3*) have shown that the subarachnoid spaces are barely discernible. Such images of the human fetal brain have lent support to laboratory findings suggesting that CSF may have a major role to play in the early development of the brain. The fetus with spina bifida has a poorly developed cortex with large ventricles occupying space into which cortical development would normally have occurred. The lack of pressure within the fluid system is evident in two ways: first, the fetal head circumference does not grow disproportionately quickly and, second, delivered infants with spina bifida have fontanelles that are found to be soft. In hydrocephalic infants, the fontanelles are bulging and tense with the pressure from the underlying brain and fluid. Thus, the hypothesis taken to laboratory studies is that fluid flow over the surface of the brain is important for the normal development of the cerebral cortex at least, if not the entire central nervous system.

4. Early-onset hydrocephalus in the H-Texas rat

The H-Texas (HTx) rat was observed as a spontaneous mutation in a normal Sprague-Dawley rat strain (Kohn *et al.*, 1981) and has since been maintained as an inbred strain. Mating unaffected males and females produces litters with an average of 40% affected pups. Problems begin at day 18 of gestation when morphological studies show fluid accumulation in the ventricles (*Figure 4*). Some controversy exists in the definition of the hydrocephalus as communicating, due to a block distal to the cerebral aqueduct, or as non-communicating, due to a constriction and blockage of the aqueduct (Jones, 1992, 1997; Jones and Bucknall, 1988; Oi *et al.*, 1996). Although our own investigations are continuing, they so far indicate that aqueduct closure is secondary to a problem with fluid flow or absorption through the subarachnoid spaces and arachnoid villi. Whether fluid flow is affected by subarachnoid or aqueduct problems, the result is that cerebrospinal fluid does not flow at a normal rate and accumulates in the lateral ventricles (*Figure 4*). To date, this model of early-onset hydrocephalus has been used to investigate the post-natal effects of raised intracranial pressure, the pathophysiological consequences of the condition and the efficacy of treatment involving fluid drainage (Boillat *et al.*, 1993, 1997; Harris *et al.*, 1994, 1996; Hawkins *et al.*, 1997; Jones *et al.*, 1995a, b, 2000; Miyan *et al.*, 1998; Miyazawa and Sato, 1991; Yoshida *et al.*, 1990). However, we have become

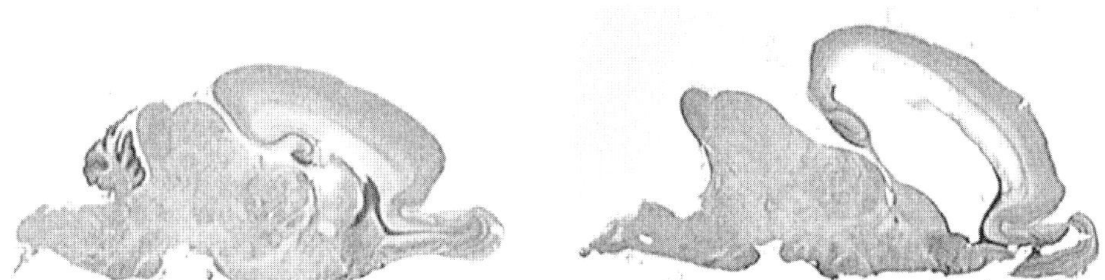

(a) (b)

Figure 4. *Low power photomicrographs of sagittal sections through a normal (a) and hydrocephalic (b) HTx rat brain. Hydrocephalus results in fluid accumulation in the ventricles and the raised pressure is associated with damage to white matter and later damage to the grey matter. (× 40 methyl green–pyronin stain.)*

increasingly aware in the HTx rat of pre-natal developmental effects of the aberrant CSF dynamics that are probably independent of raised intracranial pressure (Vetsika *et al.*, 1999) and mirror the observations of abnormal development of the brain in the spina bifida fetus described above.

From day 18 of gestation there is a steady increase in ventricular dilatation (*Figure 5*) as fluid accumulates following a change in secretion, absorption, or following a constriction and blockage in the CSF pathway. Our observations suggest that a sequence of events lead to constriction of the aqueduct that eventually leads to hydrocephalus. Studies of intracranial pressure changes in affected pups do not find a rise in pressure until 10 days after birth, 14–15 days after the onset of ventricular dilatation (Jones, 1985; Jones and Bucknall, 1987; Jones and Lopman, 1998; Kaiser and Jones, 1991). The relationship between ventricular volume and pressure is not simple and studies in humans fetuses find little correlation that could be used clinically to estimate pressure (Rekate, 1997). Thus, the fact that the ventricles enlarge does not imply pressure increase, and in the rat model raised pressure appears to be correlated with fusion of the skull plates rather than to ventricular dilatation. Together with lack of cortical development, in order to explain the observations the substance of the brain parenchyma must be subjected to a significant degree of compression and stretching as fluid accumulates prior to a rise in pressure.

The clue that something is affecting development of the HTx rat brain comes from observations of changes in the germinal matrix (Vetsika *et al.*, 1999) and from observations that apoptosis is decreased in the hydrocephalic brain compared to the normal (Miyan *et al.*, 1998). In affected HTx brains the germinal matrix is thicker, containing more cells than normal and, in addition, shows two peaks of increasing thickness not observed in normal brains over the period of rapid cell division and migration (*Figure 6*) (Vetsika *et al.*, 1999). Clearly, this implies greater proliferative activity and an increase in the production of neurons and glia to supply the cortex. However, the

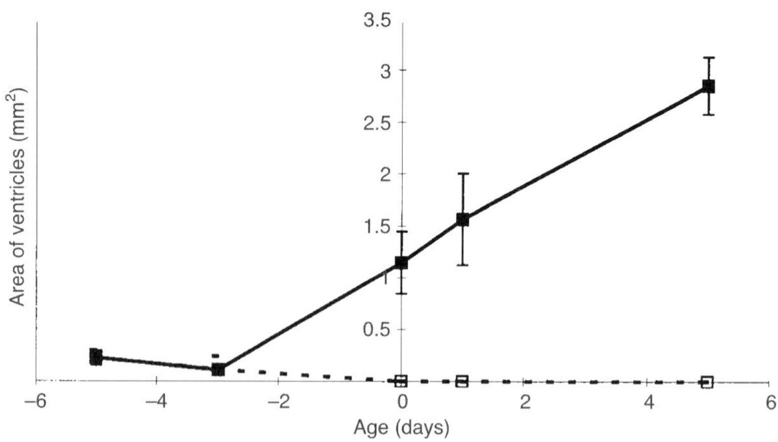

Figure 5. *Plots of ventricular areas measured from coronal sections taken at or near the optic chiasma in normal (dotted line) and hydrocephalic (solid line) rat brains (mean ± SEM).*

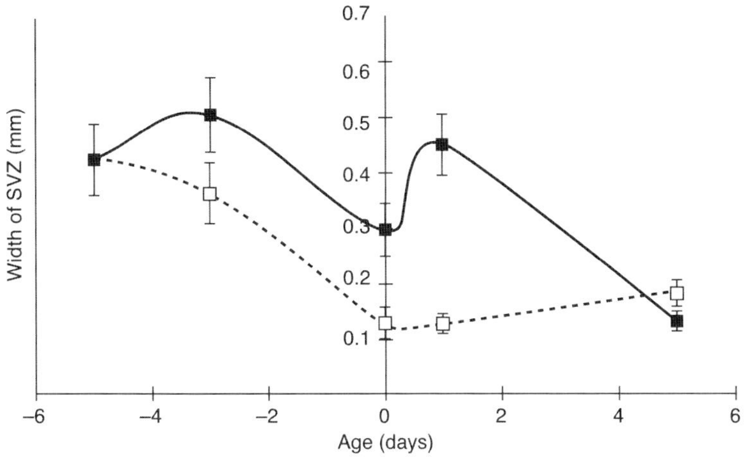

Figure 6. *Plot of measurements of germinal matrix thickness for different ages in normal (dotted line) and affected (solid line) rat fetuses and pups.*

cortex remains thin suggesting that these cells do not migrate into the cortex, migrating somewhere else (perhaps horizontally to the olfactory system) or dying within the germinal layers (Miyan *et al.*, 1998). In our recent work (Mashayekhi *et al.*, 2000) we used injections of 2-bromodeoxyuridine (BrDu) to investigate the fate of cells born in the germinal epithelium during the peak periods of proliferation inferred from morphological measurements (*Figure 7*). These experiments confirm greater germinal matrix activity in the affected HTx cortex since many more cells are labelled with BrDu than in the normal brain. In addition, the experiments show that cells born

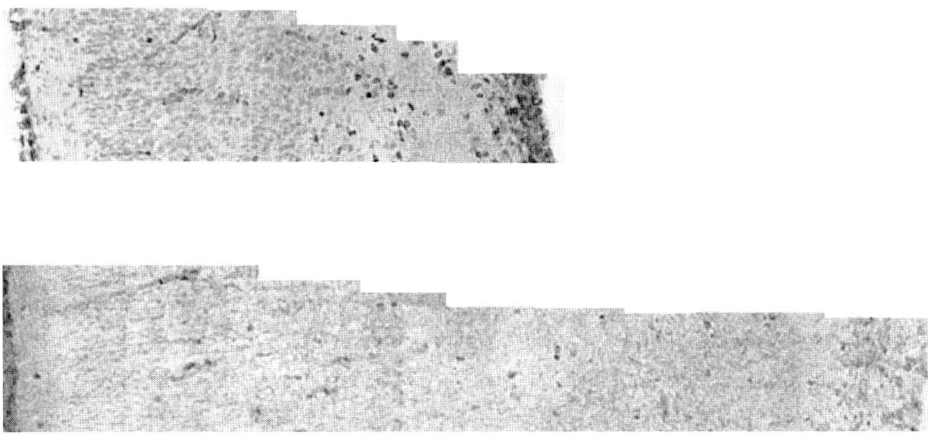

Figure 7. *BrDu-labelled cells generated the day before birth and identified 48 hours later. In the affected cortex there are more labelled cells in the germinal matrix and intermediate zone but few, if any, in the cortical plate.*

in the germinal matrix of affected brain migrate into the intermediate zone between the germinal layers and cortical plate, but appear to fail to penetrate into the cortical plate.

5. Development of the brain

The development of the brain, particularly the cerebral cortex, has received a great deal of attention over the past few years and many of the molecular mechanisms are becoming clear (see other chapters in this symposium volume). Attention has focused on two main areas of research: first, on brain stem cells, found in the neural tube and ventricular zone of the cerebral cortex, and the mechanisms involved in their proliferation and differentiation into neurones and glia (Lambert de Rouvroit and Goffinet, 1998; Marin-Padilla, 1999; Mollgard & Jacobsen, 1984; Super *et al.*, 1998). Second, on the hierarchical signals involved in the migration of neurones from the germinal epithelium into the cortex along radial glia, including the role of Cajal-Retzius (CR) cells, located in the marginal zone, the extracellular glycoprotein, reelin, secreted by these cells, and the associated receptor and intracellular signalling systems involved in the response to reelin by migrating cells (Frotscher, 1997, 1998; Gonzalez *et al.*, 1997; Marin-Padilla, 1998; Noctor *et al.*, 1999; Rice *et al.*, 1998; Soriano *et al.*, 1997; Zecevic *et al.*, 1999).

The pia has an important role in the process of corticogenesis. Firstly due to an influence on the radial glial fibres and secondly because of a trophic dependence of CR cells on the meningeal cells. Super *et al.* (1997) demonstrated that destruction of meningeal cells leads to disruption of cortical lamination and that this is due to a loss of function, and death of the underlying CR cells. They demonstrated that the 6-hydroxydopamine injection into the subarachnoid space, used to damage the meningeal cells, did not directly affect CR cells since normetanephrine (a blocker of non-neuronal uptake of catecholamines, including 6-hydroxydopamine) injected at the same time resulted in CR cell survival. This would not have occurred if the treatment directly affected CR-cells. In the developing neocortex two populations of CR-cells are recognized (Bradford *et al.*, 1977). Those only found in the developing brain that degenerate in the adult brain, and those that persist in the adult brain. The former have processes that are likely to be in direct contact with the CSF during development and subsequently are influenced by interactions with pial cells. Since the CSF has direct access to meningeal cells as well as being able to pass through into the brain parenchyma, it would be an excellent route for delivery of a signal involved in coordinating the activity of the germinal epithelium with the migration signals released from the pial surface and marginal zone and with the modulation of the properties of radial glial cells.

6. A role for cerebrospinal fluid

A number of studies have identified the cerebrospinal fluid as a carrier of important cytokines, such as transforming growth factor-beta (TGFβ) (Van Setten *et al.*, 1999), nerve growth factor (NGF) (Kasaian and Neet, 1989; Xia *et al.*, 2000), brain-derived neurotrophic factor (BDNF) (Korhonen *et al.*, 1998), neurotrophin-3 (NT-3) (Xia *et al.*, 2000), insulin-like growth factor (IGF) (Heinze *et al.*, 1998), and glial-derived neurotrophic factor (GDNF) (Ikeda *et al.*, 1999), amongst others, which are present at specific times during development or under specific physiological conditions or challenges. Changes in the concentration of some of these growth factors can lead to serious consequences of imbalance in CSF production and absorption (Johanson *et al.*, 1999; Johnson *et al.*, 1992). In addition, the CSF has the potential to act as a signalling pathway for physiological control systems since it has been demonstrated to contain molecules such as corticotropin-releasing factor, adrenocorticotropin, leptin, interleukins and leukotrienes. Moreover, CSF contains concentrations of neurotransmitters and neuropeptides that change with the physiological activity of the animal (Armstrong *et al.*, 2000; Arnold *et al.*, 1999; Azuma *et al.*, 2000; Berga *et al.*, 2000; Biou *et al.*, 2000; Giehl *et al.*, 1997; Givens *et al.*, 2000; Grimes *et al.*, 2000; Grouzmann *et al.*, 2000; Hain *et al.*, 1999; Hirai and Seki, 2000; Johanson *et al.*, 1976; Kalin *et al.*, 2000; Kraemer, 1986; Levine *et al.*, 2000; Lopez-Cortes *et al.*, 2000; Mayatepek *et al.*, 2000; Mizuno *et al.*, 2000; Riikonen *et al.*, 1999; Rodriguez *et al.*, 1999; Roldan *et al.*, 1999; Whalen *et al.*, 2000; Zeidler, 2000).

A recent study (Proescholdt *et al.*, 2000) demonstrates that compounds carried in the CSF not only circulate rapidly (seconds–minutes) through the CSF pathway, but also have fast (minutes–hours) access to most regions of the brain itself, gaining entry across the pia and ependymal layers through gap junctions, other channels and modulated junctions, and via active transport or diffusion. Moreover, compounds carried in the CSF begin to be restricted in their penetration through the brain parenchyma through development (Johanson, 1980) supporting an important role for CSF in development, since it has far easier access during these stages. Morphological and electron microscopic studies also show some of the cells underlying the ependyma (presumably neurons or glia) having processes passing between ependymal cells and in contact with the CSF. This view of CSF is supported by experiments demonstrating that intraventricular injections of growth factor can modulate the development of the cortex (Fukumitsu *et al.*, 1998; Giehl *et al.*, 1997) and can result in defective development or abnormalities (Johanson *et al.*, 1999; Moinuddin and Tada, 2000). Indeed, Moinuddin and Tada (2000) find that TGFβ is raised in patients following subarachnoid haemorrhage who later develop communicating hydrocephalus, and that injections of TGFβ into normal mice result in communicating hydrocephalus following an increase in cellularity of the leptomeninx, as well as fibrosis with a consequent narrowing of the CSF space between the meninges. The narrowing causes a disturbance in CSF flow that leads to the development of slowly progressive ventricular dilatation. The sequence may well mirror the unknown changes occurring in the affected HTx rat, since altered expression of TGFβ has been found in choroid plexus secretion from these rats (Cai *et al.*, 1999). Together, these studies indicate that the CSF is an important and rapid route for molecular signal transfer between different areas of the brain, including molecules delivered into the CSF by the choroid plexus, and that alterations in molecular signals carried in the CSF can have devastating effects on both the developing and adult brain.

Further research about how CSF is involved in development of the brain will give important new information to add to the growing picture of molecular mechanisms involved in proliferation and differentiation of pluripotent brain stem cells, and in the migration, stratification and terminal functional differentiation of neurons and glia. For hydrocephalus, identification of factors secreted by the choroid plexus and their transport to key target areas in the brain may lead to new and better methods of treating hydrocephalus, not only post-natally, but also prenatally. Restoring the CSF pathways and normalizing the CSF circulation rather than removing CSF from the ventricular system may be the way forward.

Acknowledgements

The MR scans in *Figure 3* were provided by Professor Paul Griffiths of the Academic Department of Radiology, Royal Hallamshire Hospital, University of Sheffield, UK. We have been aided by the excellent scientific and technical abilities of our colleagues and collaborators; in particular we thank Kelly Vetsika, Stephanie Girard, Jane Owen-Lynch, Nick Ritchie and Janet Wilson-Walsh. Our research is supported by The Wellcome Trust and The Royal Society.

References

Armstrong, C.S., Wuarin, L. and Ishii, D.N. (2000) Uptake of circulating insulin-like growth factor-I into the cerebrospinal fluid of normal and diabetic rats and normalization of IGF-II mRNA content in diabetic rat brain. *J. Neurosci. Res.* **59**: 649–660.

Arnold, P.M., Ma, J.Y., Citron, B.A. and Festoff, B.W. (1999) Insulin-like growth factor binding proteins in cerebrospinal fluid during human development and aging. *Biochem. Biophys. Res. Commun.* **264**: 652–656.

Azuma, H., Moriyama, T., Ikeda, H., Oshima, M., Okuno, A. and Sekiguchi, S. (2000) Analysis of soluble interleukin 6 receptor in cerebrospinal fluid in inflammatory and non-inflammatory conditions. *Cytokine* **12**: 160–164.

Bannister, C.M., Russell, S.A. and Rimmer, S. (1998) Pre-natal brain development of fetuses with a myelomeningocele. *Eur. J. Paed. Surg.* **8** (Suppl. 1): 15–17.

Berga, S.L., Loucks-Daniels, T.L., Adler, L.J., Chrousos, G.P., Cameron, J.L., Matthews, K.A. and Marcus, M.D. (2000) Cerebrospinal fluid levels of corticotropin-releasing hormone in women with functional hypothalamic amenorrhea. *Am. J. Obstet. Gynecol.* **182**: 776–781; discussion 781–784.

Biou, D., Benoist, J.F., Nguyen-Thi, C., Huong, X., Morel, P. and Marchand, M. (2000) Cerebrospinal fluid protein concentrations in children: age-related values in patients without disorders of the central nervous system. *Clin. Chem.* **46**: 399–403.

Boillat, C.A., Harris, N.G., Kaiser, G.L. and Jones, H.C. (1993) Electron microscopy of the cerebral cortex in control, hydrocephalic and shunted H-Tx rats. *Eur. J. Paed. Surg.* **3** (Suppl. 1): 30–31.

Boillat, C.A., Jones, H.C., Kaiser, G.L. and Harris, N.G. (1997) Ultrastructural changes in the deep cortical pyramidal cells of infant rats with inherited hydrocephalus and the effect of shunt treatment. *Exp. Neurol.* **147**: 377–388.

Bradford, R., Parnavelas, J.G. and Lieberman, A.R. (1977) Neurons in layer I of the developing occipital cortex of the rat. *J. Comp. Neurol.* **176**: 121–132.

Cai, X., Pattisapu, J.V., Tarnuzzer, R.W., Fernandez-Valle, C. and Gibson, J.S. (1999) TGF-beta1 expression is reduced in hydrocephalic H-Tx rat brain. *Eur. J. Paed. Surg.* **9** (Suppl. 1): 35–38.

Coucke, P., Vits, L., Van Camp, G., Serville, F., Lyonnet, S., Kenwrick, S., Rosenthal, A., Wehnert, M., Munnich, A. and Willems, P.J. (1994) Identification of a 5′ splice site mutation in intron 4 of the L1CAM gene in an X-linked hydrocephalus family. *Hum. Mol. Genet.* **3**: 671–673.

Frotscher, M. (1997) Dual role of Cajal-Retzius cells and reelin in cortical development. *Cell Tiss. Res.* **290**: 315–322.

Frotscher, M. (1998) Cajal-Retzius cells, Reelin, and the formation of layers. *Curr. Opin. Neurobiol.* **8**: 570–575.

Fukumitsu, H., Furukawa, Y., Tsusaka, M., Kinukawa, H., Nitta, A., Nomoto, H., Mima, T. and Furukawa, S. (1998) Simultaneous expression of brain-derived neurotrophic factor and neurotrophin-3 in Cajal-Retzius, subplate and ventricular progenitor cells during early development stages of the rat cerebral cortex. *Neuroscience* **84**: 115–127.

Giehl, K.M., Schacht, C.M., Yan, Q. and Mestres, P. (1997) GDNF is a trophic factor for adult rat corticospinal neurons and promotes their long-term survival after axotomy in vivo. *Eur. J. Neurosci.* **9**: 2479–2488.

Givens, T.G., Paul, R.I., Bothner, J.P., Hardwick, W.E.J. and Walsh-Kelly, C.M. (2000) Cerebrospinal fluid glucose and protein in disposition and treatment decisions. *Acad. Emerg. Med.* 7: 298–302.

Gonzalez, J.L., Russo, C.J., Goldowitz, D., Sweet, H.O., Davisson, M.T. and Walsh, C.A. (1997) Birthdate and cell marker analysis of scrambler: a novel mutation affecting cortical development with a reeler-like phenotype. *J. Neurosci.* **17**: 9204–9211.

Grimes, M.A., Cameron, J.L. and Fernstrom, J.D. (2000) Cerebrospinal fluid concentrations of tryptophan and 5-hydroxyindoleacetic acid in *Macaca mulatta*: diurnal variations and response to chronic changes in dietary protein intake. *Neurochem. Res.* **25**: 413–422.

Grouzmann, E., Borgeat, A., Fathi, M., Gaillard, R.C. and Ravussin, P. (2000) Plasma and cerebrospinal fluid concentration of neuropeptide Y, serotonin, and catecholamines in patients under propofol or isoflurane anesthesia. *Can. J. Physiol. Pharmacol.* **78**: 100–107.

Hain, R.D., Hardcastle, A., Pinkerton, C.R. and Aherne, G.W. (1999) Morphine and morphine-6-glucuronide in the plasma and cerebrospinal fluid of children. *Br. J. Clin. Pharmacol.* **48**: 37–42.

Hanigan, W.C., Morgan, A., Shaaban, A. and Bradle, P. (1991) Surgical treatment and long-term neurodevelopmental outcome for infants with idiopathic aqueductal stenosis. *Childs Nerv. Syst.* 7: 386–390.

Harris, N.G., Jones, H.C. and Patel, S. (1994) Ventricle shunting in young H-Tx rats with inherited congenital hydrocephalus: a quantitative histological study of cortical grey matter. *Childs Nerv. Syst.* **10**: 293–301; discussion 301.

Harris, N.G., Plant, H.D., Briggs, R.W. and Jones, H.C. (1996) Metabolite changes in the cerebral cortex of treated and untreated infant hydrocephalic rats studied using in vitro 31P-NMR spectroscopy. *J. Neurochem.* **67**: 2030–2038.

Hawkins, D., Bowers, T.M., Bannister, C.M. and Miyan, J.A. (1997) The functional outcome of shunting H-Tx rat pups at different ages. *Eur. J. Paed. Surg.* **7** (Suppl. 1): 31–34.

Heinze, E., Boker, M., Blum, W., Behnisch, W., Schulz, A., Urban, J. and Mauch, E. (1998) GH, IGF-I, IGFBP-3 and IGFBP-2 in cerebrospinal fluid of infants, during puberty and in adults. *Exper. Clin. Endocrinol. Diabetes* **106**: 197–202.

Hirai, K. and Seki, T. (2000) Cerebrospinal fluid somatostatin levels in febrile seizures and epilepsy in children. *Neuropeptides* **34**: 18–24.

Ikeda, T., Xia, X.Y., Xia, Y.X., Ikenoue, T. and Choi, B.H. (1999) Expression of glial cell line-derived neurotrophic factor in the brain and cerebrospinal fluid of the developing rat. *Int. J. Dev. Neurosci.* **17**: 681–691.

Johanson, C.E. (1980) Permeability and vascularity of the developing brain: cerebellum vs cerebral cortex. *Brain Res.* **190**: 3–16.

Johanson, C.E., Woodbury, D.M. and Withrow, C.D. (1976) Distribution of bicarbonate between blood and cerebrospinal fluid in the neonatal rat in metabolic acidosis and alkalosis. *Life Sci.* **19**: 691–700.

Johanson, C.E., Szmydynger-Chodobska, J., Chodobski, A., Baird, A., McMillan, P. and Stopa, E.G. (1999) Altered formation and bulk absorption of cerebrospinal fluid in FGF-2-induced hydrocephalus. *Am. J. Physiol.* **277**: R263–271.

Johnson, M.D., Gold, L.I. and Moses, H.L. (1992) Evidence for transforming growth factor-beta expression in human leptomeningeal cells and transforming growth factor-beta-like activity in human cerebrospinal fluid. *Lab. Investigation* **67**: 360–368.

Jones, H.C. (1985) Cerebrospinal fluid pressure and resistance to absorption during development in normal and hydrocephalic mutant mice. *Exp. Neurol.* **90**: 162–172.

Jones, H.C. (1992) Primary aqueduct stenosis as a cause of hydrocephalus [letter; comment]. *Childs Nerv. Syst.* **8**: 183.

Jones, H.C. (1997) Aqueduct stenosis in animal models of hydrocephalus [letter; comment] [see comments]. *Childs Nerv. Syst.* **13**: 503–504.

Jones, H.C. and Bucknall, R.M. (1987) Changes in cerebrospinal fluid pressure and outflow from the lateral ventricles during development of congenital hydrocephalus in the H-Tx rat. *Exp. Neurol.* **98**: 573–583.

Jones, H.C. and Bucknall, R.M. (1988) Inherited prenatal hydrocephalus in the H-Tx rat: a morphological study. *Neuropathol. Appl. Neurobiol.* **14**: 263–274.

Jones, H.C. and Lopman, B.A. (1998) The relation between CSF pressure and ventricular dilatation in hydrocephalic HTx rats. *Eur. J. Paed. Surg.* **8** (Suppl. 1): 55–58.

Jones, H.C., Harris, N.G., Briggs, R.W. and Williams, S.C. (1995a) Shunt treatment at two postnatal ages in hydrocephalic H-Tx rats quantified using MR imaging. *Exp. Neurol.* **133**: 144–152.

Jones, H.C., Rivera, K.M. and Harris, N.G. (1995b) Learning deficits in congenitally hydrocephalic rats and prevention by early shunt treatment. *Childs Nerv. Syst.* **11**: 655–660.

Jones, H.C., Harris, N.G., Rocca, J.R. and Andersohn, R.W. (2000) Progressive tissue injury in infantile hydrocephalus and prevention/reversal with shunt treatment. *Neurol. Res.* **22**: 89–96.

Jouet, M. and Kenwrick, S. (1995) Gene analysis of L1 neural cell adhesion molecule in prenatal diagnosis of hydrocephalus. *Lancet* **345**: 161–162.

Kaiser, G. and Jones, H.C. (1991) Cerebrospinal fluid pressure in 10-day-old rats with congenital hydrocephalus. *Eur. J. Paed. Surg.* **1** (Suppl. 1): 20–22.

Kalin, N.H., Shelton, S.E. and Davidson, R.J. (2000) Cerebrospinal fluid corticotropin-releasing hormone levels are elevated in monkeys with patterns of brain activity associated with fearful temperament. *Biol. Psych.* **47**: 579–585.

Kasaian, M.T. and Neet, K.E. (1989) Nerve growth factor in human amniotic and cerebrospinal fluid. *Biofactors* **2**: 99–104.

Kohn, D.F., Chinookoswong, N. and Chou, S.M. (1981) A new model of congenital hydrocephalus in the rat. *Acta Neuropathologica* **54**: 211–218.

Korhonen, L., Riikonen, R., Nawa, H. and Lindholm, D. (1998) Brain derived neurotrophic factor is increased in cerebrospinal fluid of children suffering from asphyxia. *Neurosci. Lett.* **240**: 151–154.

Kraemer, G.W. (1986) Causes of changes in brain noradrenaline systems and later effects on responses to social stressors in rhesus monkeys: the cascade hypothesis. *Ciba Foundation Symposium* **123**: 216–233.

Lambert de Rouvroit, C. and Goffinet, A.M. (1998) A new view of early cortical development. *Biochem. Pharmacol.* **56**: 1403–1409.

Levine, J., Panchalingam, K., Rapoport, A., Gershon, S., McClure, R.J. and Pettegrew, J.W. (2000) Increased cerebrospinal fluid glutamine levels in depressed patients. *Biol. Psych.* **47**: 586–593.

Lopez-Cortes, L.F., Marquez-Arbizu, R., Jimenez-Jimenez, L.M., Jimenez-Mejias, E., Caballero-Granado, F.J., Rey-Romero, C., Polaina, M. and Pachon, J. (2000) Cerebrospinal fluid tumor necrosis factor-alpha, interleukin-1beta, interleukin-6, and interleukin-8 as diagnostic markers of cerebrospinal fluid infection in neurosurgical patients [see comments]. *Crit. Care Med.* **28**: 215–219.

MacFarlane, J.R., Du, J.S., Pepys, M.E. *et al.* (1997) Nine novel L1 CAM mutations in families with X-linked hydrocephalus. *Hum. Mutat.* **9**: 512–518.

Marin-Padilla, M. (1998) Cajal–Retzius cells and the development of the neocortex. *Trends Neurosci.* **21**: 64–71.

Marin-Padilla, M. (1999) The development of the human cerebral cortex. A cytoarchitectonic theory. *Rev. Neurol.* **29**: 208–216.

Mashayekhi, F., Bannister, C.M. and Miyan, J.A. (2000) Possible role of CSF in a developmental abnormality associated with early-onset hydrocephalus. *Eur. J. Paed. Surg.* **10** (Suppl. 1): **10**: 39–40.

Mataro, M., Poca, M.A., Sahuquillo, J., Cuxart, A., Iborra, J., de la Calzada, M.D. and Junque, C. (2000) Cognitive changes after cerebrospinal fluid shunting in young adults with spina bifida and assumed arrested hydrocephalus. *J. Neurol. Neurosurg. Psych.* **68**: 615–621.

Mawera, G. and Asala, S.A. (1996) The function of arachnoid villi/granulations revisited. *Central African J. Med.* **42**: 281–284.

Mayatepek, E., Zelezny, R. and Hoffmann, G.F. (2000) Analysis of leukotrienes in cerebrospinal fluid of a reference population and patients with inborn errors of metabolism: further evidence for a pathognomonic profile in LTC(4)-synthesis deficiency. *Clin. Chim. Acta* **292**: 155–162.

Miyan, J.A., Khan, M.I., Kawarada, Y., Sugiyama, T. and Bannister, C.M. (1998) Cell death in the brain of the HTx rat. *Eur. J. Paed. Surg.* **8** (Suppl. 1): 43–48.

Miyazawa, T. and Sato, K. (1991) Learning disability and impairment of synaptogenesis in HTX-rats with arrested shunt-dependent hydrocephalus. *Childs Nerv. Syst.* **7**: 121–128.

Mizuno, Y., Takada, H., Urakami, K., Ihara, K., Kira, R., Suminoe, A., Ohga, S., Aoki, T. and Hara, T. (2000) Neurotrophin-3 levels in cerebrospinal fluid from children with bacterial meningitis, viral meningitis, or encephalitis. *J. Child Neurol.* **15**: 19–21.

Moinuddin, S.M. and Tada, T. (2000) Study of cerebrospinal fluid flow dynamics in TGF-beta 1 induced chronic hydrocephalic mice. *Neurol. Res.* **22**: 215–222.

Mollgard, K. and Jacobsen, M. (1984) Immunohistochemical identification of some plasma proteins in human embryonic and fetal forebrain with particular reference to the development of the neocortex. *Brain Res.* **315**: 49–63.

Nicholson, C. (1999) Signals that go with the flow. *Trends Neurosci.* **22**: 143–145.

Noctor, S.C., Palmer, S.L., Hasling, T. and Juliano, S.L. (1999) Interference with the development of early generated neocortex results in disruption of radial glia and abnormal formation of neocortical layers. *Cerebral Cortex* **9**: 121–136.

Oberbauer, R.W. (1985) The significance of morphological details for developmental outcome in infantile hydrocephalus. *Childs Nerv. Syst.* **1**: 329–336.

Oi, S., Yamada, H., Sato, O. and Matsumoto, S. (1996) Experimental models of congenital hydrocephalus and comparable clinical problems in the fetal and neonatal periods [see comments]. *Childs Nerv. Syst.* **12**: 292–302.

Proescholdt, M.G., Hutto, B., Brady, L.S. and Herkenham, M. (2000) Studies of cerebrospinal fluid flow and penetration into brain following lateral ventricle and cisterna magna injections of the tracer [^{14}C]inulin in rat. *Neuroscience* **95**: 577–592.

Rekate, H.L. (1997) Recent advances in the understanding and treatment of hydrocephalus. *Semin. Pediatr. Neurol.* **4**: 167–178.

Rice, D.S., Sheldon, M., D'Arcangelo, G., Nakajima, K., Goldowitz, D. and Curran, T. (1998) Disabled-1 acts downstream of Reelin in a signaling pathway that controls laminar organization in the mammalian brain. *Development* **125**: 3719–3729.

Riikonen, R., Somer, M. and Turpeinen, U. (1999) Low insulin-like growth factor (IGF-1) in the cerebrospinal fluid of children with progressive encephalopathy, hypsarrhythmia, and optic atrophy (PEHO) syndrome and cerebellar degeneration. *Epilepsia* **40**: 1642–1648.

Rodriguez, S., Vio, K., Wagner, C., Barria, M., Navarrete, E.H., Ramirez, V.D., Perez-Figares, J.M. and Rodriguez, E.M. (1999) Changes in the cerebrospinal-fluid monoamines in rats with an immunoneutralization of the subcommissural organ–Reissner's fiber complex by maternal delivery of antibodies. *Exp. Brain Res.* **128**: 278–290.

Roldan, A., Figueras-Aloy, J., Deulofeu, R. and Jimenez, R. (1999) Glycine and other neurotransmitter amino acids in cerebrospinal fluid in perinatal asphyxia and neonatal hypoxic-ischaemic encephalopathy [see comments]. *Acta Paediatr.* **88**: 1137–1141.

Soriano, E., Alvarado-Mallart, R.M., Dumesnil, N., Del Rio, J. and Sotelo, C. (1997) Cajal–Retzius cells regulate the radial glia phenotype in the adult and developing cerebellum and alter granule cell migration. *Neuron* **18**: 563–577.

Spertell, R.B. (1980) The response of brain to transient elevations in intraventricular pressure. *J. Neurol. Sci* **48**, 343–352.

Super, H., Martinez, A. and Soriano, E. (1997) Degeneration of Cajal–Retzius cells in the developing cerebral cortex of the mouse after ablation of meningeal cells by 6-hydroxydopamine. *Brain Res. Develop. Brain Res.* **98**: 15–20.

Super, H., Soriano, E. and Uylings, H.B. (1998) The functions of the preplate in development and evolution of the neocortex and hippocampus. *Brain Res. Rev.* **27**: 40–64.

Van Camp, G., Vits, L., Coucke, P., Lyonnet, S., Schrander-Stumpel, C., Darby, J., Holden, J., Munnich, A. and Willems, P.J. (1993) A duplication in the L1CAM gene associated with X-linked hydrocephalus. *Nature Genet.* **4**: 421–425.

Van Setten, G.B., Edstrom, L., Stibler, H., Rasmussen, S. and Schultz, G. (1999) Levels of transforming growth factor alpha (TGF-alpha) in human cerebrospinal fluid. *Int. J. Develop. Neurosci.* **17**: 131–134.

Vetsika, E.K., Bannister, C.M., Buckle, A.M. and Miyan, J.A. (1999) The effects of CSF blockage in early-onset hydrocephalus on the activity of the germinal epithelium. *Eur. J. Paediatr. Surg.* **9** (Suppl. 1): 43–44.

Whalen, M.J., Carlos, T.M., Kochanek, P.M., Wisniewski, S.R., Bell, M.J., Clark, R.S., DeKosky, S.T., Marion, D.W. and Adelson, P.D. (2000) Interleukin-8 is increased in cerebrospinal fluid of children with severe head injury [see comments]. *Crit. Care Med.* **28**: 929–934.

Xia, Y.X., Ikeda, T., Xia, X.Y. and Ikenoue, T. (2000) Differential neurotrophin levels in cerebrospinal fluid and their changes during development in newborn rat. *Neurosci. Lett.* **280**: 220–222.

Yoshida, Y., Koya, G., Tamayama, K., Kumanishi, T. and Abe, S. (1990) Histopathology of cystic cavities in the cerebral white matter of HTX rats with inherited hydrocephalus. *Neurologia Medico-chirurgica* **30**: 229–233.

Zecevic, N., Milosevic, A., Rakic, S. and Marin-Padilla, M. (1999) Early development and composition of the human primordial plexiform layer: An immunohistochemical study. *J. Comp. Neurol.* **412**: 241–254.

Zeidler, M. (2000) 14–3–3 cerebrospinal fluid protein and Creutzfeldt-Jakob disease [letter]. *Ann. Neurol.* **47**, 683–684.

9

Stem cells and regeneration in invertebrate Deuterostomes

M.C. Thorndyke, M. Patruno, W.-C. Chen and P.W. Beesley

1. Introduction

The animal kingdom is divided into Protostomes and Deuterostomes based largely upon mode of development (Holland, 2000). The Protostomes comprise the vast majority of invertebrate animals, including the widely studied arthropods, molluscs and worms. Deuterostomes include the chordates and, hence, vertebrates and mammals. For obvious reasons it is the vertebrates, and mammals in particular, that have attracted most attention and indeed most effort in stem cell research has focused on these taxa. However, the Deuterostomes also include two extant invertebrate groups, both exclusively marine, the echinoderms and the tunicates, or sea squirts.

The tunicates have attracted most interest since they provide tractable and convenient models for the study of chordate ancestry because of their unique tadpole larvae with notochord and dorsal CNS (DiGregorio and Levine 1998; Wada *et al.* 1998). Echinoderms, on the other hand, while being the darling of developmental biologists for many years have lagged behind the tunicates in terms of their potential for providing information on stem cell origins and phylogeny in deuterostomes.

2. Echinoderms

Since the work of Boveri (see Balzer, 1967 for review) and then others (Davidson *et al.*, 1998; Horstadius, 1973), sea urchin embryos and larvae have had a long and distinguished history as models for the study of development. Echinoderms are unique in the context of deuterostome developmental mechanisms and process because their classic larval stage is planktonic with a clear bilateral symmetry, and it is this well-ordered developmental programme, from zygote through blastula, to gastrula and then larva, that has proved to be so attractive to developmental biologists. This larva is transformed

Brain Stem Cells, edited by J. Miyan, M. Thorndyke, P.W. Beesley and C. Bannister.

at metamorphosis into an adult with a quite different pentaradiate body plan (Peterson *et al.*, 2000). Of particular interest here is the idea that in those species showing maximal indirect development, that is with a long planktonic larval stage, groups of so-called set-aside cells are present which retain pluripotentiality and provide the stem cells from which the adult body is formed (Peterson *et al.*, 1997). Indeed, there are good recent examples showing that such larvae are capable of extensive regeneration (Vickery and McClintock, 1998; Vickery *et al.*, 2001). Adult echinoderms are well known for their regenerative potential and here too there are a number of recent studies which indicate the presence of a pluripotent and dynamic stem cell population in the adult (Candia Carnevali *et al.*, 1998; Moss *et al.*, 1998a). The most extensively studied classes are the crinoids (feather stars) and asteriods (sea stars)

2.1 *Regeneration in crinoids*

Crinoid regeneration has been studied extensively in *Antedon mediterranea* (Candia Carnevali *et al.*, 1993, 1995, 1997, 1998). In this model, new structures originate from the recruitment of migratory, proliferating cells that are morphologically undifferentiated and produced at the level of the brachial nerve (amoebocytes) and the coelomic epithelium (coelomocytes). These migrating cells produce a blastema at the site of tissue loss and the main mechanism of regeneration here seems to be based on epimorphosis.

The sites of cell proliferation include the blastema itself, the entire length of the coelomic epithelium and in the brachial nerve. Candia Carnevali *et al.* (1995, 1997) suggest that migratory amoebocytes give rise to blastemal cells while coelomocytes are important in originating coelomic cell types. In the brachial nerve other migratory cells are also present, including phagocytes and granule cells. The early regenerative blastema consists primarily of undifferentiated amoebocytes. By 5–7 days post ablation (p.a.) the regenerative process is well underway and substantial growth and differentiation of new tissues is taking place and a well developed regenerating arm is formed together with lateral pinnules within one week. These growth processes are supported by extensive cell-cycle activity as evidenced by studies using bromodeoxyuridine (BrdU) (Candia Carnevali *et al.*, 1995, 1997, 1998) that show labelling of blastemal amoebocytes and in the epithelial layers of the coelomic compartments.

Neurogenesis in crinoid regeneration. During normal embryonic development, neurogenesis is initiated when a small section of the embryonic ectoderm adopts a prospective neural fate. In vertebrates, mesodermally derived neural-inducing factors such as chordin and noggin act as functional antagonists to the bone morphogenetic family of growth regulators (BMPs) and determine the area of neural competence (Christian, 2000). In *Drosophila,* two similar genes play a pivotal role in neurogenesis: *decapentaplegic* (*Dpp*) and *short gastrulation* (*sog*), the homologues of vertebrate BMP4 and chordin, respectively. In addition to regulating the distinction between neural and non-neural ectoderm and organizing mediolateral patterning in the neural plate, vertebrate BMP2 and BMP4 are equally involved in early neural crest development. Here, modified levels of BMP2/BMP4 at the junction between neural and non-neural ectoderm initiate neural crest formation. Thus an early and highly conserved event in neurogenesis is the expression of BMPs and their antagonists as a means to establish areas of neural potential and commitment. Recent work in our laboratory has been addressing the idea that a similar cascade might be involved in echinoderm regeneration.

2.2 *BMPs in echinoderm regeneration*

BMP homologues have been identified in sea urchin embryos and it is clear that they are important for early developmental events, as indicated by expression patterns of the *univin* gene in *Strongylocentrotus purpuratus* (Stenzel *et al.*, 1994). We have explored the idea that these molecules may also be important in adult regeneration and involved in cell lineage regulation. We used RT-PCR to characterize a BMP homologue from the crinoid *Antedon bifida* and have identified a molecule belonging to the BMP subfamily (Patruno, 2001; Thorndyke *et al.*, 2000). High amino acid similarity scores were found with sea urchin BMP2/4 homologues such as *TgBMP2/4* (82%) (Hwang *et al.*, 1999), as well as human BMP-2 (81%) and BMP-4 (78%) respectively (*Figure 1*).

Wholemount *in situ* hybridization showed that expression of *AnBMP2/4* occurs rapidly following amputation with the first hybridization signal detectable at the beginning of the typical repair phase (20 hr post ablation) with a high level of transcription along the oral side of the regenerating arm in association with the coelomic canal epithelium. Expression remained high and spatially restricted to the coelomic epithelium over the following hours of regeneration. At early regenerative

```
1 AYYCEGECPFPISDHLNATNHAIVQTLVNSVNPQLVPKACCVPTDLSSISILYLDENEKVVLKVYKEMVVEGCGCR
2 ....Q......LV..........V......A....A..A.....D..P..M.....SDA.I..N.QN.I.VS..---
3 ....H......LAE...T.....V......VN.A....A.....E..A..M.....Y......N.QD.........
4 ....R......LAE...T.....V......VN.A....A.....E..A..M.....Y......N.QD........GRT
5 .F..H......LA....S.....V......VN-SKI..A.....E..A..M............N.QD.........--
6 .F..H.D....LA....S.....V......VN-SSI..A.....E..A..M.....YD.....N.Q..........--
7 .F..H......LP....S.....V......VN-SNI..A..I..E..P..L.....Y...I..N.QD........L--
8 ....Q.D....LS....G.....V......VN.AA...A..V..Q..S..M..M..VNN....N.QD.M.V.....--
9 ....D......LGER..G.....I......IDNRA...V..A..K..G..M..F.N..N...RQ.ED....A....--
```

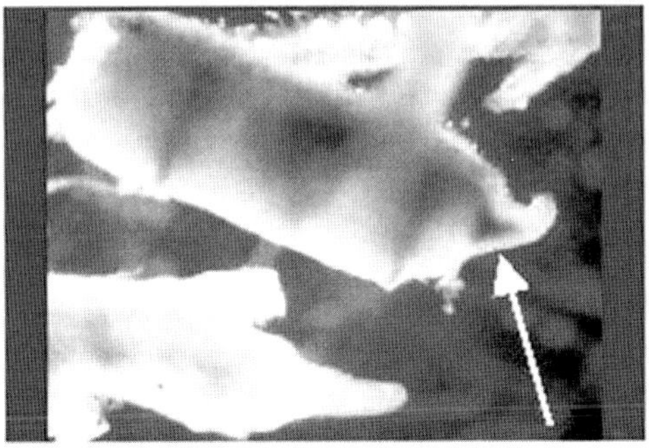

Figure 1. *Top, alignment of vertebrate and invertebrate BMP2/4 homologues and the gene portion of* A. bifida *(top lane) using Clustal W. Asterisks and dots within the sequences indicate amino acid identity. Dots in the bottom line indicate amino acid similarities. The highest degree (82%) of similarity is shown with the* T. gratilla *BMP2/4 homologue; the lowest (67%) with the* S. purpuratus univin. *GenBank accession numbers are: 2)* A. rubens *BMP 2/4, CAB63584; 3)* T. gratilla *TGBMP2/4, AF133305; 4)* S. purpuratus *BMP2/4, AF119713; 5)* H. sapiens *BMP-2; 6)* H. sapiens *BMP-4, BAA06410.1; 7)* D. rerio *BMP2-4, AAC60286; 8)* J. coenia, *AAB46367; 9)* S. purpuratus-*UNIVIN, P48970. Bottom,* in situ *hybridization preparation for* AnBMP2/4 *at one week p.a. in* A. bifida. *Clear evidence of expression may be seen in the blastemal region (arrow). The sense control (below) is negative.*

stages (48–72 hr pa), the amputated stump comprises a well-formed blastema which grows, as noted above, due to the active contribution of migrating, proliferating stem cells derived from both the brachial nerve and the coelomic canals (Candia Carnevali *et al.*, 1997). Here, a marked *AnBMP2/4* expression can be seen in the blastema. At 1 week post ablation (pa), when the regenerating arm is well defined, the pattern of expression shifts slightly to become proximally located in the blastemal cells alone (*Figure 1*), while the signal in the coelomic canals fades. At 2 weeks pa the signal is no longer detectable by *in situ* hybridization at a time when tissue identities are well established and morphogenetic growth is taking place (Candia Carnevali *et al.*, 1995). It is tempting to speculate, then, that *AnBMP2/4* plays an important part in the regulation of cell identity and lineage specification during regeneration. In this way, *AnBMP2/4* expression during the early phases of regeneration, when the fates of migrating cells are being determined, might play a key role in specification events.

2.3 *Regeneration in asteroids*

In asteroid regeneration, no discrete blastema as a centre of cell proliferation is evident. Here, morphallaxis (the regenerative process where the new cells arise by proliferation from existing tissues) (*Figure 2*) is the mechanism employed whereby most cell-cycle activity is concentrated in the epidermal layer of the radial nerve cord and in the epithelium of the coelomic compartments. Migratory wandering coelomocytes are also associated with the regenerating structures, but seem to be involved only to a minor extent (Moss *et al.*, 1998a). The two different populations of cells,

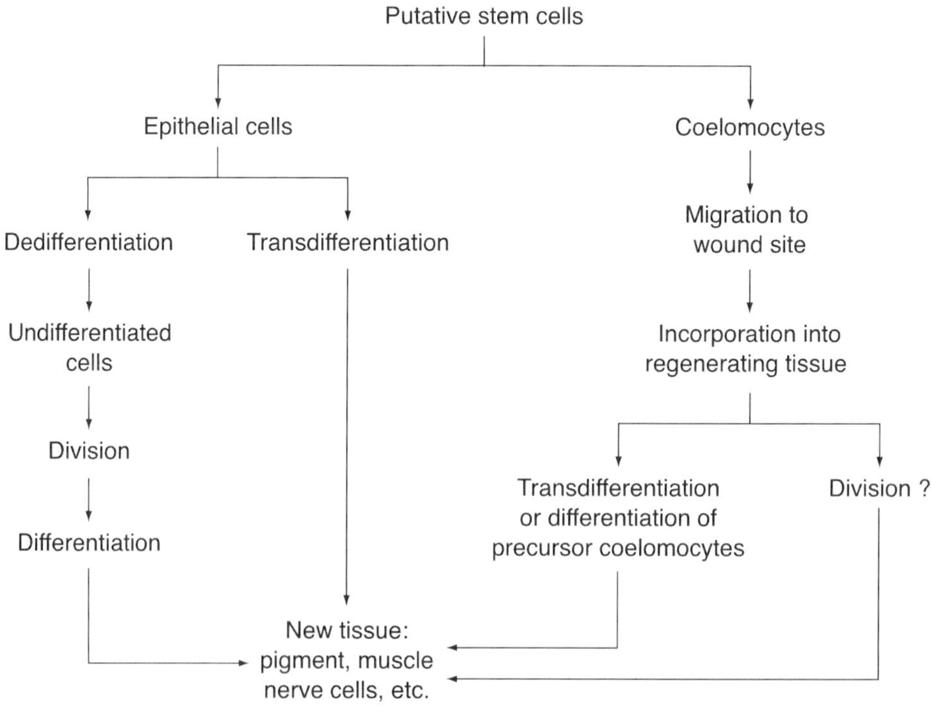

Figure 2. Potential lineage pathways for putative stem cells in asteroid regeneration.

epidermal and coelomic mesothelial, provide the major contribution in terms of development/regrowth of tissues. The morphallactic process in asteroids is rather slow when compared to the epimorphosis of crinoids. In *Asterias* most cell-cycle activity, detected using BrdU, occurs between 7 and 21 days, a period when the arm tip begins to undergo elongation. It is also at this period that the regenerating tip begins to change shape and adopt the more familiar adult organization with distal optic cushion and terminal tentacle. Findings here suggest (a) de-differentiation, proliferation and re-differentiation and/or (b) trans-differentiation of committed cells, which must switch from their 'default' pathway to alternative pathways specifying other cell types, including neurons (Moss *et al.*, 1998a).

2.4 *The role of* Hox *genes in neuronal regeneration in echinoderms*

Homeobox-containing genes comprise a network of transcription factors implicated in the regulation of body patterning in animal development (Prince, 1998; Reichert and Simeone, 1999; Veraksa *et al.*, 2000). The homeobox is a 180 bp, protein-coding DNA sequence originally identified in the *Drosophila* homeotic selector genes *Antennapedia* and *Ultrabithorax*, and the segmentation gene *fushi tarazu* (McGinnis *et al.*, 1984; Scott and Weiner, 1984). The *Hox* complex of homeobox-containing genes has attracted considerable attention because of their role in patterning the nervous system during development (Chan and Yan, 1999; DiGregorio and Levine, 1998; Prince, 1998).

A characteristic feature of the *Hox* gene cluster is one of structural colinearity, that is the spatial and temporal pattern of expression of each *Hox* gene reflects its spatial position in the cluster. Thus, genes at the 3′ end of the cluster are expressed in the more anterior compartments of the embryo, while genes at the 5′ end of the cluster have an overlapping expression with more posterior compartments. In addition, the temporal expression of each *Hox* gene corresponds to its position in the cluster, 3′ genes being expressed earliest. The phenomenon of colinearity and the involvement of *Hox* genes in patterning of the nervous system is illustrated *par excellence* in the spatiotemporal expression of *Hox* genes in the rhombomeres of the developing vertebrate hindbrain (Keynes and Krumlauf, 1994). Here, *Hox* genes are expressed in overlapping domains along the A–P axis of the developing hindbrain and spinal cord and deficiency in the expression of any *Hox* gene causes malformation of the corresponding CNS compartment (Maconochie *et al.*, 1996). *Hox* gene nomenclature reflects their location in the cluster and their expression pattern along the A–P axis. Thus, vertebrate *Hox* genes 1–3, 4–8 and 9–13 are assigned to anterior, medial and posterior groups, respectively. Thus, although *Hox* genes are not involved in determination of stem-cell fate, they are of crucial importance in specifying the positional fate of these cells along the A–P axis of the developing animal. *Hox* genes are also strongly implicated in regulation of the positional fate of cells in regenerating animals.

The role of *Hox* genes in regeneration has been explored in both both Protostomes and Deuterostomes, with attention focusing in particular on planarians and amphibians. In both groups, a number of *Hox* genes have been successfully isolated from regenerating blastemas by use of PCR amplification of cDNA and cDNA libraries (Bartels *et al.*, 1993; Bayascas *et al.*, 1997, 1998a, b; Gardiner *et al.*, 1995; Orii *et al.*, 1999; Savard and Tremblay, 1995; Savard *et al.*, 1988). For example, in axolotl limbs, *Hoxd-8* and *Hoxd-10* are detectable in the stump mesenchyme within 24 hours of amputation (Torok *et al.*, 1998). Moreover, the expression pattern of *Hox* and homeobox genes in regeneration is not simply a reiteration of the developmental programme. For example, the newt homeobox gene *NvHbox2* (the homeodomain is identical to that of mouse *Hox-5.5*) is expressed in the blastema of the regenerating limb as a transcript of 3.4 kb that is not detectable in the normal limb (Brown and Brockes, 1991). *Hoxc-10* has been suggested to be uniquely expressed in regenerating axolotl forelimbs (Carlson *et al.*, 2001).

Unlike vertebrates, echinoderms possess only a single *Hox* gene complex and the organization of the *Hox* gene cluster of the echinoid *Strongylocentrus pupuratus* is illustrated in *Figure 3*.

Recent studies in our laboratory have focused on cloning *Hox* genes from the sea star *Asterias rubens* and investigating their role in regeneration of the radial nerve cord following arm tip ablation. PCR, using degenerate *Hox* primers followed by cloning and sequence analysis has led to the identification of six *A. rubens Hox*-type genes and two other genes more closely related to the *Gbx* and *Gax/Mox* family than the *Hox* gene cluster. The characteristic sequences for homeobox positions 60–135 and derived amino acid sequences for these eight *A. rubens* genes are illustrated in *Figure 4*. The putative homeodomain encoded by each of these genes shows a high degree of similarity with their counterparts in amphioxus, *Drosophila* and other echinoderms, such as *Asterina minor*.

We have analysed the expression pattern of the *A. rubens Hox1* gene in regenerating arm tips at various stages of regeneration by RT-PCR of arm tip RNA extracts and by western blotting using an antibody specific for the *Asterina minor Hox1* gene (gift of Dr. T. Mito and Professor K. Endo, Tokyo University) (*Figure 5*).

PCR analyis shows that *ARHox1* expression is up-regulated throughout the period of regeneration compared to control animals (*Figure 5*). Whole-mount immuno-cytochemistry shows that *ARHox1*-like reaction product is first detected at 10 days post ablation (i.e. following the initial 5–7 day period of wound healing). At this stage the labelling is confined to the area immediately proximal to the ablation surface (*Figure 5*). Although PCR reveals that *ARHox1* is expressed in the non-regenerating animal this is probably related to the growth and plasticity of this structure in the normal adult, the up-regulation of this gene during regeneration reflecting the increased cell proliferation and migration observed in the arm tip throughout this phenomenon. Thus, the available evidence strongly suggests that expression of *Hox* genes early after commitment of stem cells to a neuronal lineage is instrumental in specifying the positional fate of these cells during asteroid arm regeneration.

Further evidence for the crucial role of *Hox1* in determining the positional fate of cells in echinoderms comes from studies of *Hox1* expression during larval metamorphosis in *Asterina minor*. *AMHox1* shows an intriguing pattern of expression during larval metamorphosis. Here, the bilateral larval symmetry with a well-defined A–P axis is transformed into a radial asymmetry, due to atrophy of structures on the right-hand side of

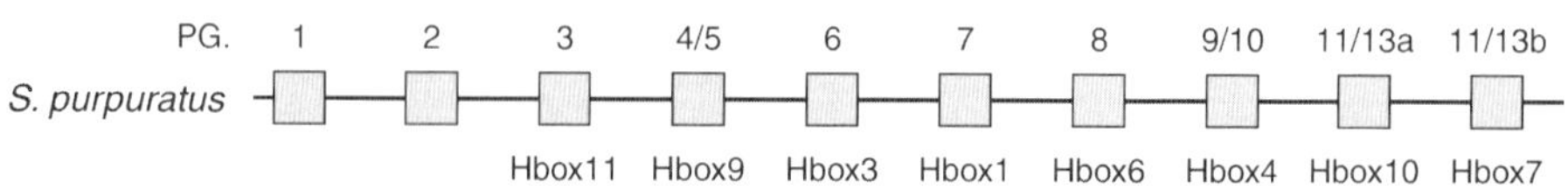

Figure 3. *Organization of the echinoid Hox gene cluster (*Strongylocentrotus purpuratus*). The names of the echinoid Hox genes found in earlier literature (bottom) and their paralogous affinities with vertebrate Hox genes (top).*

```
ARHox1
CACTTCAACAAGTATCTTACGCGGGCACGGCGAGTCGAAATCGCCACCATGTTGGGTTTGAACGAAACACAAGTC   1
H  F  N  K  Y  L  T  R  A  R  R  V  E  I  A  T  M  L  G  L  N  E  T  Q  V

ARHox4/5
CATTTCAACCGCTACCTCACAAGACGGCGTCGCATCGAGATCGCTCATGCGCTCGGTCTTACCGAACGCCAAATC   6
H  F  N  R  Y  L  T  R  R  R  R  I  E  I  A  H  A  L  G  L  T  E  R  Q  I

ARHox6
CACTTCAATCGGTACCTTACCCGTAGGAGACGCATCGAGATCGCCCAGTCACTCGGTCTCTCGGAGAGGCAGATT   1
H  F  N  R  Y  L  T  R  R  R  R  I  E  I  A  Q  S  L  G  L  S  E  R  Q  I

ARHox7
CACTACAACCGTTACCTCACAAGAAGAAGACGGATCGAGCTGTCGCATCAACTCGGACTCACAGAAAGACAGATC   3
H  Y  N  R  Y  L  T  R  R  R  R  I  E  L  S  H  Q  L  G  L  T  E  R  Q  I

ARHox9/10
CTCTTTAACATGTATCTAACCCGAGACAGACGAGTGGAGATTGCCAGGCAGCTCAATCTGACAGAGAGACAGGTC   2
L  F  N  M  Y  L  T  R  D  R  R  V  E  I  A  R  Q  L  N  L  T  E  R  Q  V

AR-HboxP9
CAGTCACATCAGTATCTCACAAGAGACCGGAGAGCGAGACTAGCCCAGACCCTCACACTAACGGAGAGACAAGTC   3
Q  S  H  Q  Y  L  T  R  D  R  R  A  R  L  A  Q  T  L  T  L  T  E  R  Q  V

AR-Mox
AGCCACAATAACTACCTTACGCGCCTGCGCAGATACGAAATAGCAGTCACGCTCAACTTGACAGAAAGACAGGTA   4
S  H  N  N  Y  L  T  R  L  R  R  Y  E  I  A  V  T  L  N  L  T  E  R  Q  V

AR-Gbx
CACAGTAAGAAATATCTGACTCTAAGCGAGAGATCGCAGATCGCCAGGGCTCTTCATCTTAGTGAAGTCCAGGTT   30
H  S  K  K  Y  L  T  L  S  E  R  S  Q  I  A  R  A  L  H  L  S  E  V  Q  V
```

Figure 4. *Nucleotide sequences (homeobox positions 60–135) and derived amino acid sequences (homeodomain positions 21–45) of homeobox fragments from* Asterias rubens. *Numbers at right indicate the quantity of inserts sequenced.*

the body. Subsequently, the metamorphosing larva adopts the pentaradiate symmetry of the adult structure. In early stages of metamorphosis *AMHox1* expression is detected as a continuous stellate pattern. However, during the later stages of metamorphosis *AMHox1* immunoreactivity is confined to five patches of cells located at the base of each of the five arms. These results suggest a key role for selected *Hox* genes in specifying the pentaradiate symmetry characteristic of the adult echinoderm (*Figure 6*).

In conclusion, we now have good evidence to suggest that both BMP homologues and *Hox* genes play key roles during regeneration in echinoderms. However, early markers for cells committing to a neuronal lineage are still proving elusive.

3. Tunicates

As invertebrate members of the phylum Chordata, tunicates (particularly ascidians or sea squirts) have figured prominently in the exploration of potential origins of the sophisticated vertebrate CNS (Meinertzhagen and Okamura, 2001; Satoh, 1994; Wada *et al.*, 1998). The tunicate tadpole larva is arguably the archetypical ancestral chordate since it provides us with most, if not all, of the requirements deemed essential for such an organism including notochord, muscular post-anal tail and dorsal CNS. At larval settlement and metamorphosis most of the chordate characters exhibited by the larva regress and the neural complex of the adult comprises an epithelial glandular structure (the neural gland) in association with a simple cerebral ganglion (*Figure 7*). This simple adult neural complex shows little homology with the chordate CNS, although for many years attempts have been made to equate it with the vertebrate pituitary based partly on morphology and partly on the expression of neuropeptides including GnRH (Bollner *et al.*, 1997).

The adult neural complex is unusual because in several species, including the widely studied *Ciona intestinalis*, it has the potential to regenerate in its entirety following

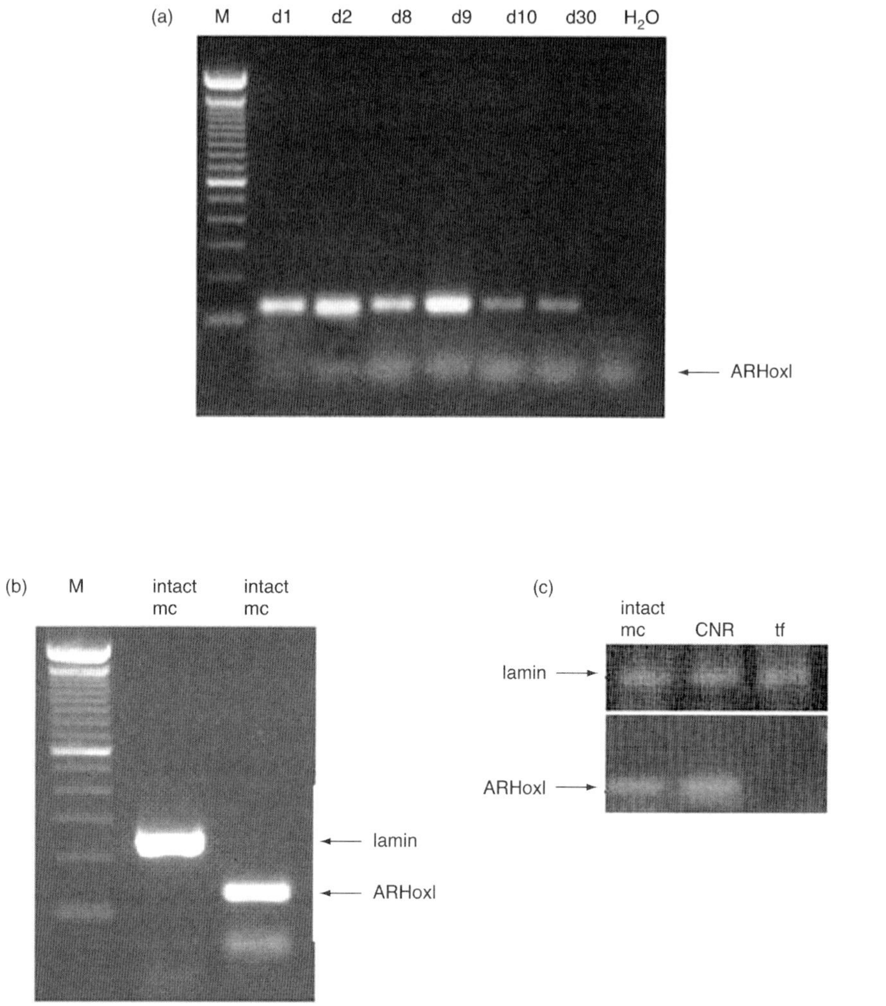

Figure 5. *Analysis of* Asterias rubens *Hox1 gene expression by RT-PCR. As a control, the lamin gene, a ubiquitous cellular transcript, was also amplified. (a) PCR amplification showing ARHox1 expression in regenerating radial nerve cords from 1 day to 30 days post ablation. (b) ARHox1 expression is detected in intact radial nerve cords (rnc) suggests that this gene is permanently expressed. (c) PCR amplification showing ARHox1 expression in* A. rubens *nervous system (CNR: circumoral nerve ring) but not in other tissue such as tube feet (tf).*

complete ablation. The regenerated complex resembles the normal adult structure in all respects (although slightly smaller), including the ordered distribution of specific neurotransmitter cell types. Thus, the regenerated neurons express detectable amounts of peptide transmitters/modulators (CCK, substance P, GnRH) and of the classical transmitter GABA by 14 and 21 days post ablation, respectively (Bollner *et al.*, 1992, 1993a, b, 1995, 1997) (*Figure 8*).

The new neurons are derived from at least two populations of putative stem cells. The first population gives rise to the large neurons that become localized to the cortical rind of the ganglion and are born prior to ablation. Their most likely source is

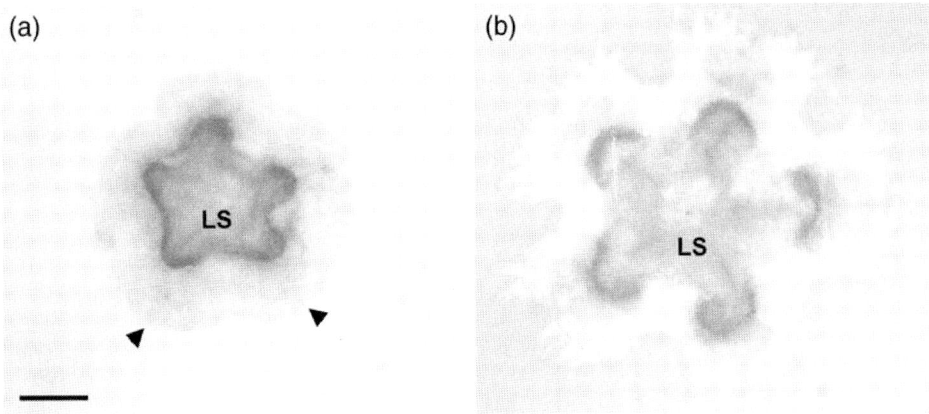

Figure 6. *Expression pattern of* Asterina minor Hox1 *in 6.5 day (a) and 8 day (b) old juveniles as revealed by immunocytochemistry (see text) LS, left somatocele. Scale bar 200 μm.*

pluripotent haemocytes (Bollner *et al.*, 1993a, b). Accumulation of blood cells at the site of regeneration is observed within 24 h of ablation of the neural complex. This initial accumulation of cells represents a wound-healing rather than a regenerative response. However, the work of Tanaka (Tanaka *et al.*, 1999) provides evidence for a direct link between wound healing and regeneration in the newt limb. Critical to this regenerative capacity is the ability of cells to lose their differentiated characteristics and re-enter the cell cycle. Cultured newt myotubes are able to re-enter the cell cycle following serum stimulation (Tanaka and Brockes, 1998). This process is mediated by thrombin cleavage of serum components to generate a ligand that acts directly on myotubes to stimulate their passage through the S-phase of the cell cycle. The downstream target protein, which leads to re-entry of the myotubes into the cell cycle, is phosphorylated retinoblastoma protein (Tanaka *et al.*, 1997). Thus, in the newt limb the ability of cells to de-differentiate and re-enter the cell cycle is critical to the regenerative process.

The second population of neuronal precursors is born post ablation and gives rise to small and medium sized neurons, scattered throughout the ganglion. BrdU labelling

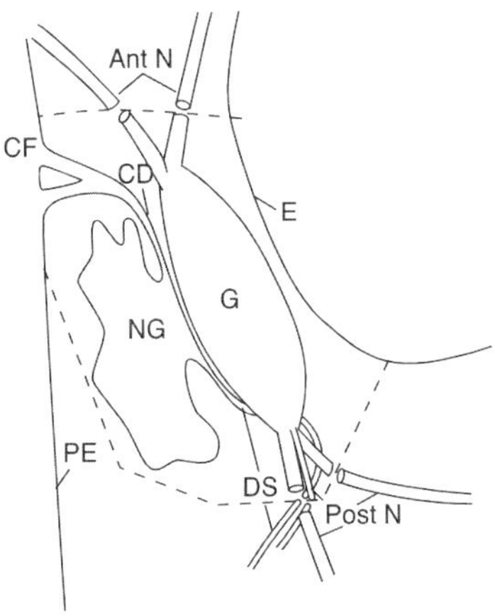

Figure 7. *Diagram to illustrate the neural complex in* Ciona intestinalis *and the ablation area (dashed lines). Ant N, anterior nerve roots; CD, ciliated duct; CF, ciliated funnel; DS, dorsal strand; E, epidermis; G, ganglion; NG, neural gland; PE, pharyngeal epithelium; Post N, posterior nerve roots. From Bollner* et al. *1992.*

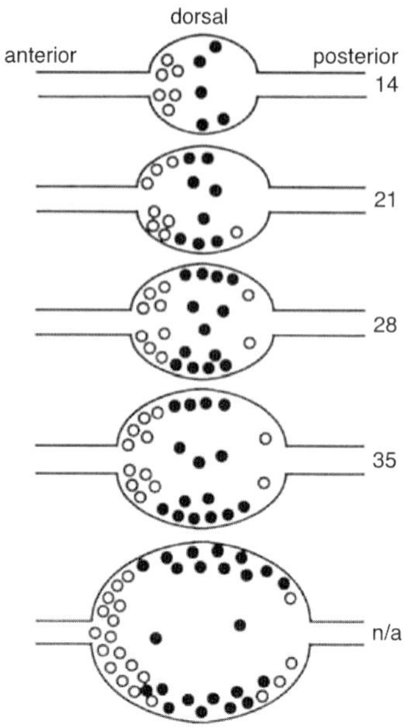

Figure 8. Illustration showing the distribution of Substance P- (closed circles) and cholecystokinin- (open circles) immunoreactive neurons during regenration of the neural ganglion in Ciona intestinalis. *Days post-ablation are noted on the right. At 14 days CCK-Li neurons preferentially locate to the anterior while SP-Li neurons remain more central. As regeneration progresses SP-Li neurons increase in number and become concentrated in the ganglion cortex with a few scattered in the neuropile. CCK-Li neurons also expand in number and remain concentrated anteriorly close to the nerve trunk exit points. na, normal adult. From Bollner* et al. *1992.*

shows that these cells are born principally within the first 14 days post ablation. The most likely source of these cells is the dorsal strand epithelium, a specialized epithelial structure that runs posteriorly from the neural gland to the gonads. Thus, as in echinoderms, both differentiation and transdifferentiation appear to contribute to the process of regeneration (Bollner *et al.*, 1993a, b). Immunocytochemical studies with anti-GnRH antibodies and BrdU suggest that immunoreactive cells present in the dorsal strand plexus and some neurons in the regenerated ganglion arise by delamination from the dorsal strand epithelium (*Figure 9*), providing further evidence that this structure provides one source of precursor cells (Bollner *et al.*, 1997). This finding is particularly relevant on two counts: first, it recalls the neurogenic mechanism seen both in neural placodes and neural crest (Baker and Bronner-Fraser, 1997) and second, in budding tunicates (see below) epithelial delamination is similarly involved as a source of adult stem cells (Manni *et al.*, 1999).

As with echinoderms noted above, a major problem in the unequivocal identification of the stem cells that give rise to new neurons during regeneration is the lack of suitable markers for localization of these cells at their site of generation. Production of monoclonal antibodies (mAbs) against neural tissue from *Ciona intestinalis* has resulted in production of two mAb markers, Cin 1 and Cin 2, that recognize all, or a subpopulation of, ganglion neurons at early stages of regeneration (Bollner *et al.*,

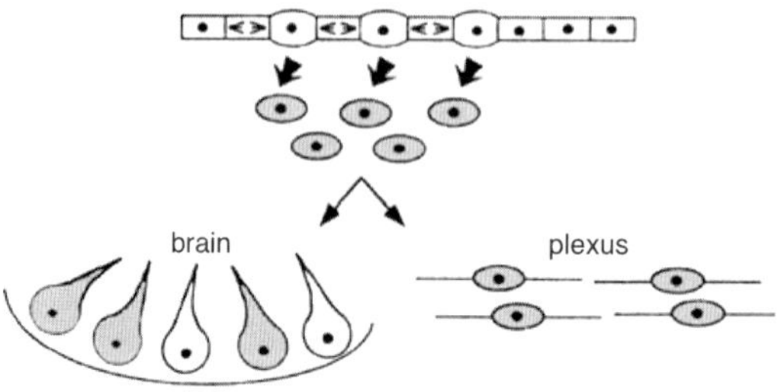

Figure 9. Schematic showing putative lineage pathway of dorsal strand-derived stem cells. "Neuroblasts" (shaded) delaminate from the dorsal strand to be recruited either as GnRH-expressing neurones in the brain or in the dorsal strand plexus. From Bollner et al. *1997.*

1995). Immunocytochemical studies using Cin 1 demonstrate the presence of neuronal cells that are extending processes at the site of regeneration as early as three days post ablation. A sub-population of Cin 2 immunoreactive neurons can be distinguished as early as five days post ablation. However, neither of these markers is useful for detection of stem cells at around the time of commitment to a neuronal lineage. The characterization of native early adult neural markers for regeneration in both ascidians and echinoderms must be a high priority for future work.

4. Perspectives

Such echinoderm and ascidian models offer distinct advantages of simplicity and experimental tractability, while at the same time as Deuterostomes they have clear value for our understanding of the origins and phylogeny of chordate neurogenesis. However, as noted above, one of the problems associated with current work in both echinoderms and ascidians is the difficulty associated with the unequivocal identification of cell lineage and the *precise* nature and origins of the pluripotent stem cells involved. Thus, while we have good evidence for the role of BMPs in early boundary definition and for *Hox* genes in the later stages of positional regulation, we have yet to identify markers for early neuronal commitment.

Downstream of BMP's role in the regulation of ectodermal neural/non-neural identity, a number of other genes and gene cascades are involved in the further specification of neuronal fate. These genes, too, show a high degree of conservation across the animal kingdom (Chan and Jan 1999; Christian 2000). For example, suppression of BMP2/BMP4 in the neural ectoderm and the adoption of neural fate leads to the activation of the bHLH proneural and neurogenic genes. These promote neural fate and establish the proneural fields (Chan and Jan, 1999). In *Drosophila*, proneural genes include *atonal* and *achaete-scute*, while vertebrate homologues are *MASH/XASH/MATH* and the *neurogenins* (*ngn*1 and *ngn*2). Within the proneural fields, lateral inhibition then operates to restrict neural fate. Here, the Delta and Notch pathways interact such that elevated expression of Delta in one cell will activate the Notch inhibitory pathway in adjacent cells and restrict their potential to adopt a neural fate (Lendahl, 1998). Downstream of Delta, neuronal differentiation genes such as *neuro-D* operate to confirm neuronal identity. In parallel, neural crest delamination and migration depend upon the expression of *Wnt* and *FgF* genes (Baker and Bronner-Fraser, 1997). Clearly, in exploring the complete picture of stem cell lineage and the part it plays in echinoderm and tunicate regeneration it will be important to develop expression markers for such native gene cascades likely to be involved in the establishment of neural identity. Development of *in situ* hybridization probes for echinoderm and ascidian homologues should prove to be of seminal importance for identification of the stem cell populations that give rise to the new adult neurons.

5. Functional studies

While functional studies have yet to be carried out on the regenerating systems described here, it is worth noting that microinjection of mRNAs in *S. purpuratus* embryos confirmed that BMP signalling operates in the embryonic environment, and suggested that BMP2/4 might be important in establishing morphogenetic gradients in echinoderm by regulating the position of the ectoderm/endoderm boundary (Angerer and Angerer, 2000; Angerer *et al.*, 2000). In order to take the study of both echinoderm and tunicate neuronal regeneration forward it will clearly be important to develop experimental systems where predictions based on established gene expression profiles might be tested.

It is conceivable that colonial ascidians may provide some advantage as experimental systems, as shown by work on the role of retinoic acid in the regulation of budding (Kamimura *et al.*, 2000). On the other hand, *in vitro* models are becoming available and in our laboratory we have developed a robust system for the culture of both echinoderm and tunicate neural explants (Moss *et al.*, 1998b). It is likely that these will prove valuable tools with which to investigate both the role of gene expression and potential stem cell identities in regeneration. For example, pre-labelling putative stem cells such as haemocytes with cell markers like DiI will enable us to explore key sequences of neuronal lineage gene expression in parallel with a mechanism which allows us to follow original cell identity.

Acknowledgements

The authors would like to acknowledge the support of BBSRC and NERC for much of their work. In addition, support is gratefully acknowledged from NATO, a Thomas Holloway Studentship (MP), RHUL Strategic Investment Fund (MCT) and the University of London Central Research Fund (MCT). Thanks also to Michael and Minako Vickery for access to their data prior to publication.

References

Angerer, L.M. and Angerer, R.C. (2000) Animal–vegetal axis patterning mechanisms in the early sea urchin embryo. *Dev. Biol.* **218**: 1–12.

Angerer, L.M., Oleksyn, Logan, C.Y., McClay, D.R., Dale, L. and Angerer, R.C. (2000) A BMP pathway regulates cell fate allocation along the sea urchin animal–vegetal embryonic axis. *Development* **127**: 1105–1114.

Baker, C. and Bronner-Fraser, M. (1997) The origins of the neural crest. Part II: an evolutionary perspective. *Mechs. Develop.* **69**: 13–29.

Balzer, E. (1967) Theodor Boveri: Life and work of a Great Biologist. University of California Press, Berkeley and Los Angeles.

Bartels, J.L., Murtha, M.T. and Duddle, F.H. (1993) Multiple *Hox/HOM*-class homeoboxes in Platyhelminthes. *Mol. Phylogenet. Evol.* **2**: 143–151.

Bayascas, J.R., Castillo, E., Muonz-Marmol, A.M. and Salo, E. (1997) Planarian *Hox* gene: novel patterns of expression during regeneration. *Development* **124**: 144–148.

Bayascas, J.R., Castillo, E., Munoz-Marmol, A.M., Baguna, J. and Salo E. (1998a) Synchronous and early activation of planarian *Hox* genes and the re-specification of body axes during regeneration in *Dugesia (G.) tigrina* (Turbellaria; Tricladida). *Hydrobiologia* **383**: 125–130.

Bayascas, J.R., Castillo, E. and Salo, E. (1998b) Platyhelminthes have a hox code differentially activated during regeneration, with genes closely related to those of spiralian protostomes. *Dev. Genes Evol.* **208**(8): 467–473.

Bollner, T., Beesley, P. and Thorndyke, M.C. (1992) The pattern of substance P- and cholecystokinin-like immunoreactivity during the regeneration of the neural complex in the ascidian *Ciona intestinalis*. *J. Comp. Neurol.* **325**: 572.

Bollner, T., Beesley, P.W. and Thorndyke, M.C. (1993a) Distribution of GABA-like immunoreactivity during post-metamorphic development and regeneration of the central nervous system in the ascidian *Ciona intestinalis*. *Cell Tiss. Res.* **272**: 553.

Bollner, T., Beesley, P.W. and Thorndyke, M.C. (1993b) Substance P-like and cholecystokinin-like immunoreactivity during post-metamorphic development of the central nervous system in the ascidian *Ciona intestinalis*. *Cell Tiss. Res.* **272**: 545.

Bollner, T., Howatt, S., Thorndyke, M.C. and Beesley, P.W. (1995) Regeneration and post-metamorphic development of the central nervous system of the protochordate *Ciona intestinalis*: a study with monoclonal antibodies. *Cell Tiss. Res.* **279**: 421–432.

Bollner, T., Beesley, P.W. and Thorndyke, M.C. (1997) Investigation of the contribution from peripheral GnRH-like immunoreactive 'neuroblasts' to the regenerating central nervous system in the protochordate *Ciona intestinalis. Proc. R. Soc. Lond. B* 264: 1117–1124.

Brown, R. and Brockes, J.P. (1991) Identification and expression of a regeneration-specific homeobox gene in the newt limb blastema. *Development* **111**: 489–496.

Candia Carnevali, M.D., Bonasoro, F., Lucca, E. and Thorndyke, M.C. (1995) Pattern of cell proliferation in the early stages of arm regeneration in the feather star *Antedon mediterranea*. *J. Exp. Zool.* **272**: 464–474.

Candia Carnevali, M.D., Bonasoro, F. and Biale, A. (1997) Pattern of bromodeoxyuridine incorporation in the advanced stages of arm regeneration in the feather star *Antedon mediterranea. Cell Tiss. Res.* **289**: 363–374.

Candia Carnevali, M.D., Bonasoro, F., Patruno, M. and Thorndyke, M.C. (1998) Cellular and molecular mechanisms of arm regeneration in crinoid echinoderms: the potential of arm explant. *Dev. Genes Evol.* **208**: 421–430.

Candia Carnevali, M.D., Lucca, E. and Bonasoro, F. (1993) Mechanism of arm regeneration in the feather star *Antedon mediterranea*: healing of wound and early stages of development. *J. Exp. Zool.* **267**: 299–317.

Carlson, M.R., Komine, Y., Bryant, S.V. and Gardiner, D.M. (2001) Expression of Hoxb13 and Hoxc10 in developing and regenerating axolotl limbs and tails. *Dev. Biol.* **229**(2): 396–406.

Chan, Y.-M. and Jan, Y.N. (1999) Conservation of neurogenic genes and mechanisms. *Curr. Opin. Neurobiol.* **9**: 582–588.

Christian, J.L. (2000) BMP, Wnt and Hedgehog signals: how far can they go? *Curr. Opin. Cell Biol.* **12**: 211–219.

Davidson, E.H., Cameron, R.A. and Ransick, A. (1998) Specification of cell fate in the sea urchin embryo: summary and some proposed mechanisms. *Development* **125**: 3269–3290.

Di Gregorio, A. and Levine, M. (1998) Ascidian embryogenesis and the origins of the chordate body plan. *Curr. Opin. Genet. Develop.* **8**: 457–463.

Gardiner, D.M., Blumberg, B., Komine, Y. and Bryant, S.V. (1995) Regulation of HoxA expression in developing and regenerating axolotl limbs. *Development* **121**: 1731–1741.

Holland, L.Z. (2000) Body-plan evolution in the Bilateria: early antero-posterior patterning and the deuterostome-protostome dichotomy. *Curr. Opin. Genet. Develop.* **10**: 434–442.

Horstadius, S. (1973) Experimental Embryology of Echinoderms. Oxford; Clarendon Press pp 192.

Hwang, S.P.L., Chen, C.A. and Chen. C.P. (1999) Sea urchin TgBMP2/4 gene encoding a bone morphogenetic protein closely related to vertebrate BMP2 and BMP4 with maximal expression at the later stages of embryonic development. *Biochem. Biophys. Res. Comm.* **258:** 457–463.

Kamimura, M., Fujiwara, S., Kawamura, K. and Yubisui, T. (2000) Functional retinoid receptors in budding ascidians. *Develop. Growth Diff.* **42**: 1–8.

Keynes, R. and Krumlauf, R. (1994) Hox genes and regionalization of the nervous system. *Annu. Rev. Neurosci.* **17**: 109–132.

Lendahl, U. (1998) A growing family of Notch ligands. *BioEssays* **20**: 103–107.

Maconochie, M.S. and Nonchev *et al.* (1996). Paralogous Hox genes: functions and regulation. *Annu. Rev. Genet.* **30**: 529–556.

Manni, L., Lane, N.J., Sorrentino, M., Zaniolo, G. and Burighel, P. (1999) Mechanism of neurogenesis during the embryonic development of a Tunicate. *J. Comp. Neurol.* **412**: 527–541.

McGinnis, W., Garber, R.L., Wirz, J., Kuroiwa, A. and Gehring, W.J. (1984) A homologous protein-coding sequence in Drosophila homeotic genes and its conservation in other metazoans. *Cell* 37: 403–408.

Meinertzhagen I.A. and Okamura Y. (2001) The larval ascidian nervous system: the chordate brain from its small beginnings. *TINS* **24**: 401–411.

Moss, C., Hunter, A.J. and Thorndyke, M.C. (1998a) Patterns of bromodeoxyuridine incorporation and neuropeptide immunoreactivity in the regenerating arm of the starfish *Asterias rubens*. *Phil. Trans. Royal Soc. UK* **353**: 421–436.

Moss, C., Beesley, P.W., Thorndyke, M.C. and Bollner, T. (1998b) Preliminary observations on ascidian and echinoderm neurons and neural explants *in vitro*. *Tissue Cell* **30**: 517–524.

Orii, H., Kato, K. *et al.* (1999) The planarian HOM/HOX homeobox genes (Plox) expressed along the anteroposterior axis. *Dev. Biol.* **210**: 456–468.

Patruno M. 2001 The role of stress proteins and the Transforming Growth Factor-beta family (TGF-βs) in echinoderm regeneration. Ph.D. Thesis University of London, UK.

Peterson, K.J., Cameron, R.A. and Davidson, E.H. (1997) Set-aside cells in maximal indirect development: evolutionary and developmental significance. *BioEssays* **19**: 623–631.

Peterson, K.J., Arenas-Mena, C. and Davidson, E.H. (2000) The A/P axis in echinoderm ontogeny and evolution: Evidence from fossils and molecules. *Evol. Develop.* **2**: 93–101.

Prince, V.E. (1998) Hox genes and segmental patterning of the vertebrate hindbrain. *Am. Zool.* **38**: 634–646.

Reichert, H. and Simeone, A. (1999) Conserved usage of gap and homeotic genes in patterning the CNS. *Curr. Opin. Neurobiol.* **9**: 589–595.

Satoh, N. (1994) *Developmental Biology of Ascidians*. Cambridge University Press, Cambridge, pp. 234.

Savard, P. and Tremblay, M. (1995) Differential regulation of *HoxC6* in the appendages of adult urodeles and anurans. *J. Mol. Biol.* **249**: 879–889.

Savard, P., Gates, P. B. *et al.* (1988) Position dependent expression of a homeobox gene transcript in relation to amphibian limb regeneration. *EMBO J.* **7**(13): 4275–4282.

Scott, M.P. and Weiner, A.J. (1984) Structural relationships among genes that control development: Sequence homology between the *Antennapedia, Ultrabithorax*, and *Fushi tarazu*. *Proc. Natl Acad. Sci. USA* **81**: 4115–4119.

Stenzel, P., Angerer, L.M., Smith, B.J., Angerer, R.C. and Vale, W.W. (1994) The univin gene encodes a member of the transforming growth factor-β superfamily with restricted expression in the sea urchin embryo. *Dev. Biol.* **166**: 149–158.

Tanaka, E.M. and Brockes, J.P. (1998) A target of thrombin activation promotes cell cycle re-entry by urodele muscle cells. *Wound Repair Regen*. **6**: 371–381.

Tanaka, E.M., Gann, A.A., Gates, P.B. and Brockes, J.P. (1997) Newt myotubes reenter the cell cycle by phosphorylation of the retinoblastoma protein. *J. Cell Biol.* **136**: 155–165.

Tanaka, E.M., Dreschel, D.N. and Brockes, J.P. (1999) Thrombin regulates S-phase re-entry by cultured newt myotubes. *Curr. Biol.* **9**: 792–799.

Thorndyke M.C., Patruno M., Chen, W-C., Beesley, P.W., Graham, A., McGonnell, I. and Candia Carnevali M.D. (2000) Expression studies on homeobox and BMP homologues in echinoderm neuronal regeneration. *Soc. Neurosci. Abstr.* 116–119.

Torok, M.A., Gardiner, D.M. *et al.* (1998) Expression of HoxD genes in developing and regenerating axolotl limbs. *Dev. Biol.* **200**(2): 225–233.

Veraksa, A., Del Campo, M.and McGinnis, W. (2000) Developmental patterning genes and their conserved functions: from model organisms to humans. *Mol. Genet. Metab.* **69**: 85–100.

Vickery, S.M. and McClintock, J.B. (1998) Regeneration in Metazoan larvae. *Nature* **394**: 140.

Vickery, M.C.L., Vickery, M.S., McClintock, J.B. and Amsler, C.D. (2001) Utilization of a novel deuterostome model for the study of regeneration genetics: molecular cloning of genes that are differentially expressed during the early stages of larval se star regeneration. *Gene* (in press).

Wada, H., Saiga, H., Satoh, N. and Holland, P.W.H. (1998) Tripartite organisation of the ancestral chordate brain and the antiquity of placodes: insights from ascidian *Pax-2/5/8, Hox* and *Otx* genes. *Development* **125**: 1113–1122.

Index